MUSHROOMS AND OTHER FUNGI

MUSHROOMS AND OTHER FUNGI

Aurel Dermek

Cover illustration: Moira Grice

ARCO PUBLISHING, INC.
New York

A NOTE ON MEASUREMENTS AND QUANTITIES

In accordance with established botanical practice, measurements are given in metric, but as a guide to those who are not familiar with the metric system an *approximate* equivalent in the customary or imperial system is added in parentheses. For measurements in micrometres (0.001mm) no equivalent is practicable.

For recipes quantities are given in metric and imperial measures, and also in American standard measures. The proportions have been adjusted for each column: follow any but do not mix measurements from more than one column.

COMMON NAMES

Where possible common names are given for the fungi described in this book, but for some species no common names exist.

Published 1985 by Arco Publishing, Inc.
215 Park Avenue South, New York, NY 10003

English language edition, designed and produced by
Autumn Publishing Limited, Chichester, England

Library of Congress Catalog Card Number: 84-71484
ISBN 0-668-06304-1

Printed in Czechoslovakia

PREFACE

The name fungus (plural fungi) is given to a large group of simple plants which contain no green colour. It includes moulds, smuts, rusts, mildews, and yeasts, as well as the larger fungi which grow in woods and meadows. This volume is concerned with the larger edible fungi, commonly called mushrooms, and with their poisonous relatives, often known as toadstools.

Fungi have been known to Man since ancient times, and have formed part of his diet, along with fruit, vegetables and meat. With experience he learned to differentiate between one fungus and another, and to recognise edible and poisonous species. He began to classify them according to their flavour, and naturally concentrated on those with the most pleasing taste. Up to the first quarter of the 20th century, the gathering and selling of edible fungi was a seasonal source of income in the rural areas of many European countries.

Although times have changed, the interest in fungi has not diminished. The growing pressures of civilization have brought with them a general urge to return to nature, especially as a recreation in the increased spare time technology has brought us. Many people are not just content to spend time wandering in woodlands or by the banks of rivers and lakes, just for the fun of it: they take a deeper and more practical interest in natural history. In this respect there is a growing interest in fungi, identifying and collecting them.

A small, popular book of this kind cannot give a complete and exhaustive account of the world of fungi, but in a brief form it gives some basic facts about these remarkable organisms. It concentrates especially on some selected species which the practical mushroom-picker can find in the wild. Although the description of each fungus is brief, it is sufficient for the identification and description of species.

We believe the book will provide a good guide to mushroom-pickers, whose rambles in search of fungi can extend from early spring to late in the fall. It can help beginners to identify several dozen species of edible and poisonous fungi. A word of warning here: if in doubt always consult an experienced mycologist (fungus expert), since some poisonous specimens are not easy to distinguish from edible varieties.

Finally, we hope that this book may inspire readers to make a deeper study of fungi. So leading them towards more specialized literature on the subject.

GENERAL DESCRIPTION

Fungi are organisms that differ from green plants not only in their appearance but also in their chemical composition, their development and their way of life. For this reason, they form a separate part of the plant kingdom, and some modern botanists even group them into a separate kingdom on their own.

With the aid of chlorophyll, the green colouring matter, green plants can make their own food out of minerals and water using the energy of sunshine to assimilate this food and thus build their bodies. The structure of a green plant consists of roots, stems, and leaves. Fungi contain no chlorophyll, so they cannot manufacture their own food. They take the nourishment they need from green plants or from animals. For this reason they grow mainly in woodland, where there is plenty of dead organic material on which to feed.

To most people the word 'fungus' means the well-developed fruiting body of a sizeable mushroom, which catches the eye by its size, shape and colour. The body of the fungus proper is formed by a dense network of branching filaments (*hyphae*, singular *hypha*); this mass is known as the *mycelium*. The part of the mushroom we see is merely the fruiting body, the reproductive part of the fungus, which produces spores. Fungi mainly reproduce asexually by means of these spores, but a few species are known in which reproduction is sexual; however, the way in which this occurs has not yet been fully studied.

Fungi can be found everywhere. They are most frequent in woods, meadows and fields, and in parks and gardens. But they can also be found on piles of wood, on dumps, in mines, rivers, the sea and in the home. According to current research pileate (capped) fungi are most common in the wooded parts of the northern hemisphere. Detailed research has yet to be carried out on the fungi of the tropical forests and the woodlands of the southern hemisphere.

Some species of fungi can be found worldwide, for example the Chanterelle (*Cantharellus cibarius*), the Fairy ring Mushroom (*Marasmius oreades*), the Oyster Mushroom (*Pleurotus ostreatus*), the Fly Agaric (*Amanita muscaria*), Slippery Jack (*Suillus luteus*), the Honey Fungus (*Armillariella mellea*), the Field Mushroom (*Agaricus campestris*), the puff-ball species (*Lycoperdon perlatum*). Most of these plants grow in clearly defined, specific areas. This is particularly true of mycorrhizal fungi, those associated with the roots of certain higher plants, on which they depend for life.

So far about 40,000 species of fungi have been described. This number is still far from final, because about a thousand new species are identified every year. Outside Europe, North America and Japan there are many regions of the world where extensive mycological research is needed, and those scientists who claim that there are at least 100,000 species of fungi are probably correct.

Most species belong to the inferior fungi, the *micromycetes*, so named because they produce either small fruiting bodies or none at all. Of the superior fungi, or *macromycetes*, producing large fruiting bodies, there are approximately 15,000 species. The terms 'inferior fungi' and 'superior fungi' have no precise definition. Among the superior fungi are found the majority of the Basidiomycetes, and some Ascomycetes which have large and conspicuous fruiting bodies.

Way of Life and Importance in Nature

Fungi are saprophytic, parasitic or symbiotic. Saprophytes are decomposers; they break down and derive their nourishment from dead animal and vegetable matter; parasites feed on the living bodies of plants or animals without benefitting them in return. Some species begin life as saprophytes, but sometimes their mycelium penetrates into live plants — most often trees — and the fungi begin to live on them. These are called saproparasitic fungi.

Some fungi rely on individual species of green plants, with which they form a parasitic or symbiotic relationship. This mutual association of fungi with higher plants is called *mycorrhiza*. The hyphae of a mycorrhizal fungus either wrap themselves around fine roots of trees, shrubs or herbs or penetrate them to take all the nourishment the fungus needs. The fungal mycelium of a symbiont in return gives back to the green plants certain substances that they need to help their growth. This symbiosis is a very important feature of mycorrhiza in forestry, especially in the establishment of young trees.

Fungi grow on various kinds of substrate. They are most often found on the soil or on the dead or live parts of trees and bushes. Some kinds live as parasites on other fungi: for example, *Xerocomus parasiticus* derives its nourishment from the living fruiting bodies of the common earthball, *Scleroderma citrinum.*

Besides the nourishment derived from organic materials, fungi need suitable temperature and moisture conditions for growth. Long research indicates that fruiting bodies grow best in calm weather. Light is not important: many fungi grow in the dark. Other factors still to be discovered influence the development of fruiting bodies.

The mycelium grows in the substrate in all directions. Its length of life varies: while the mycelium of some fungi lives for only one year, that of others may live for several decades. The mycelium of many kinds of mushrooms spreads outwards, producing circles on the circumference of which the fruiting bodies are formed. The circle grows bigger every year. These are the so-called fairy rings, often seen as lush, dark green bands in short grass.

The most important function of fungi in nature is as decomposers, breaking down and destroying the dead remains of plants and animals. Together with bacteria fungi break down organic materials into simple minerals for re-use. Without this important function life on Earth would come to a halt: organic debris would heap up and there would be a lack of raw material for new life.

Man has been using fungi for centuries, mainly for the preparation of food and drink, though at first he did not realise that the raising of dough for bread or the fermentation of wine and beer was caused by yeast fungi. With the development of science people have been able to make more deliberate use of the properties of fungi in various branches of industrial production. The processing of some modern foodstuffs would be impossible without the exploitation of yeasts and moulds. Medical research has discovered the curative effects of some kinds of moulds, such as penicillin, and the pharmaceutical industry now produces various fungal-based antibiotics.

Unfortunately, not all fungi are beneficial. Many parasitic fungi cause enormous damage to crops and in forestry. Wood saprophytes, through their decomposing activities, damage timber stores, pit props, rail sleepers and the fabric of buildings, sometimes causing complete destruction. Various moulds (Micromycetes) destroy books, sculptures, paintings and other works of art. People and animals suffer from unpleasant skin diseases of fungal origin, such as thrush and athlete's foot.

Some people worry unnecessarily about the harmful qualities of certain fungi. They shun delicious edible varieties just because some are poisonous, yet there are many other kinds of plants that are not good to eat and nobody thinks of avoiding all fruit and vegetables. It is better to cultivate a rational, unprejudiced approach, learning how to tell harmless varieties from noxious ones. Fungi in their various forms contribute substantially to human diet, so it is up to us to learn how to use them beneficially.

Morphology and Anatomy of Fruiting Bodies

The fruiting bodies of fungi vary widely in shape, surface, size, texture and colour. Figure 1 shows just some of these shapes.

Practical mushroom-pickers are mainly interested in pileate (capped) fungi because they include most of the edible varieties. Later we shall pay special attention to mushrooms consisting only of a cap and a stipe (stalk). See Figure 2.

The cap is the most important part of a mushroom. The underside is the hymenophore which bears the hymenium containing the spore-producing bodies. In identifying individual species, begin by observing the shape, surface, colour and size of the cap.

The shape of the cap may be spherical, hemispherical, egg-shaped, cylindrical, conical, bell-shaped, convex, like a cushion, flat or funnel shaped (see Figure 3).

The surface of the cap may be smooth, dimpled, rough, wrinkled, streaked, bare, velvety, finely hairy or shaggy, fibrous, covered with scales, slimy, sticky or dry, dull or lustrous.

The margin of the cap may be regular, wavy or incurved, turned up or down, with a sharp or rounded edge, with or without fragments of *veil* (see below).

The colour of the cap varies — all shades of all colours may be found.

The size of the cap varies. While many species of *Marasmius* and *Collybia* are only 12-20mm (½-¾in) in diameter, some fungi are nearly 200mm (8in) across. The cap of the parasol mushroom (*Macrolepiota procera*) often tapers upwards to more than 300mm (1ft).

Besides these features, note whether or not the cap changes colour in moist conditions and whether the skin can be peeled off easily, is hard to remove, or cannot be removed at all.

The hymenophore of the cap is composed of either veins, spines, gills or tubes of varying colours (see Figure 4).

Veins are thick or thin, dense or widely spaced, single or branched, often forming a network.

Spines may be short or long, thick or thin, sharp or blunt ended, crowded together or scattered, brittle or flexible.

Gills can be wide or narrow, numerous or sparse, delicate or coarse, running from the cap margin to the stipe in various ways (see Figure 5). An important indicator is the thin edge of the gills, and whether a gill is compact, irregular, scalloped, toothed, saw-like or scaly. The gills may be hard, soft, flexible or brittle to the touch. The colour is very important, ranging through white, cream, ochre, pink, flesh-coloured, yellow, blue, green, purple, pale brown, dark brown and black. In some genera the original colour of the gills changes as they ripen, as in species of *Volvariella, Entoloma, Agaricus, Cortinarius* and *Hypholoma* among others. When young the gills of some species of *Cortinarius* are green, purple, yellow, cream, or ochre; later the ripe spores change the colour of the gills into brown or rust-brown.

Tubes may be short or long, separated from the cap flesh or detached from it, closely attached, having a notched tip, free, withdrawing or extending down the stem. The colour is usually white, yellow, ochre, yellow-green, olive, yellow-brown, or grey-pink. Generally the colour changes gradually during ripening. In some species it also changes on exposure to air. The mouths of the tubes are called pores: size, shape

a
b
c
d
e
f
g
h
i
j

and colour are the first features to note, as the pore colour can change at different stages of development and does not always match the rest of the tube. When bruised the pores of some fungi show blue, brown or brown-pink flecks.

The stipe (stalk) grows and supports the cap during the development of the fruiting body. It may be centred, off-centred or lateral according to how it is attached to the cap. Note first its shape, size, surface, and colour.

The form of the stipe may be barrel-shaped, bulging in the middle, club-shaped, cylindrical, spindle-shaped, or conical (see Figure 6). The shape of the base is important; this may be blunt, rounded, conical, bulbous, or prolonged in the form of a root.

The surface of the stipe can be smooth, grooved, dimpled, ribbed, covered by a network of fibres, bare, velvety or covered by flakes or scales, slimy, sticky or dry, dull or shiny. The stipe may be of a single shade or brightly multicoloured.

The interior of the stipe may be solid or hollow. The flesh may be hard, soft, crumbling, watery, leathery, flexible, brittle, etc.

The veil: During the development of some species of cap fungi it is very pronounced (see Figure 7). There are two types, universal and partial.

The Universal veil covers the whole fruiting body of a young fungus. While this is growing the veil gradually tears, leaving fragments of various forms on the cap and the base of the stipe.

The Partial veil connects the margin of the cap with the stipe, and in young fungi it also covers the gills or tubes. Later the veil breaks and may leave fragments hanging down at the margin forming teeth or a web. It also leaves a membranous ring on the stipe.

The flesh of fruiting bodies is very important in identification. Note its consistency, colour, and how it changes colour on exposure to air. Equally important is the taste of the 'milk' secreted by the flesh and the gills.

Spore prints are important indicators identified with the naked eye. Spores are obtained from ripe fungi by cutting the stipe off close to the gills, tubes or spines, and then laying the cap, hymanophore down, on a sheet of white paper. After a few hours a fine film of spores is deposited on the paper. Its colour is characteristic for identifying the genus. You need to have enough spores, and to determine their colour while they are fresh, as they often change shade as they dry.

The body of superior fungi consists of *hyphae*. Hyphae are fibres with interior crosswalls 2-30mm (⅛-⅞) thick, arranged in many different ways. The mycelium and the fruiting body are both composed of hyphae.

Figure 1. Different types of fruiting bodies of superior fungi: a — bracket-fungus, b — resupinate (upside-down), c — morel-like, d — club-shaped, e — shrub-like or coral-fungus, f — spherical, g — stellate (star-shaped), h — shell-like, i — pileate (having a cap), j — cup-like.

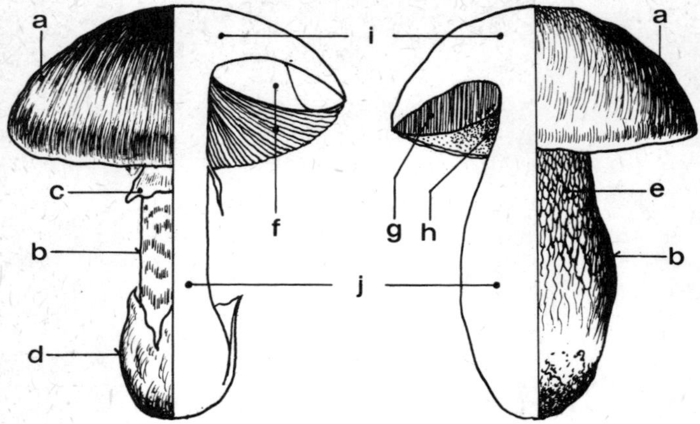

Figure 2. Parts of pileate fruiting bodies: a — cap, b — stipe, c — ring, d — volva, e — network on the stipe, f — gills, g — tubes, h — pores (mouths of the tubes), i — flesh of the cap, j — flesh of the stipe.

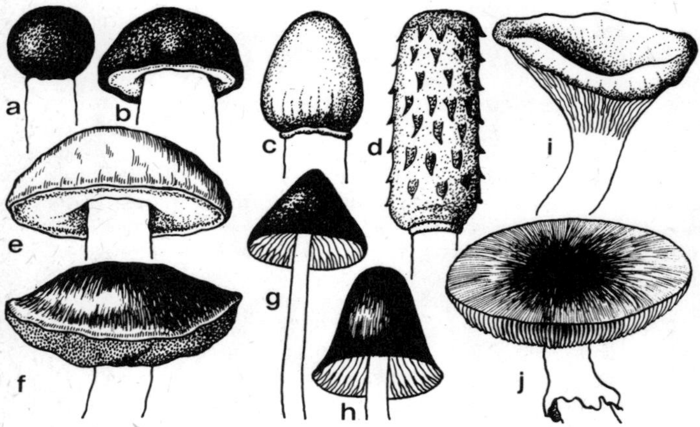

Figure 3. Different shapes of cap: a — spherical, b — hemispherical, c — egg-shaped, d — cylindrical, e — convex, f — cushion-like, g — conical, h — bell-shaped, i — funnel-shaped or concave, j — flat.

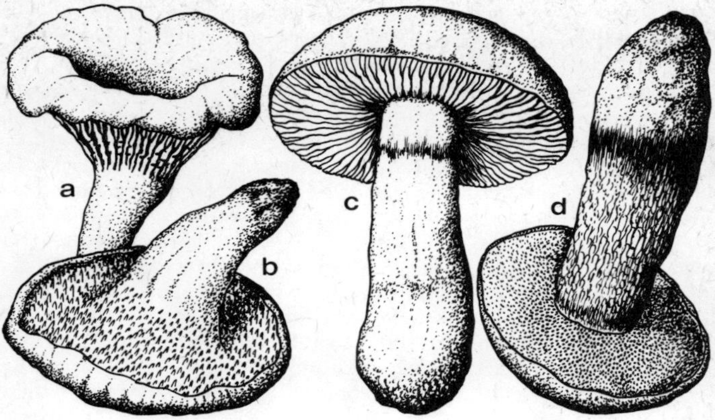

Figure 4. Different types of hymenophore: a — veins, b — spines, c — gills, d — tubes.

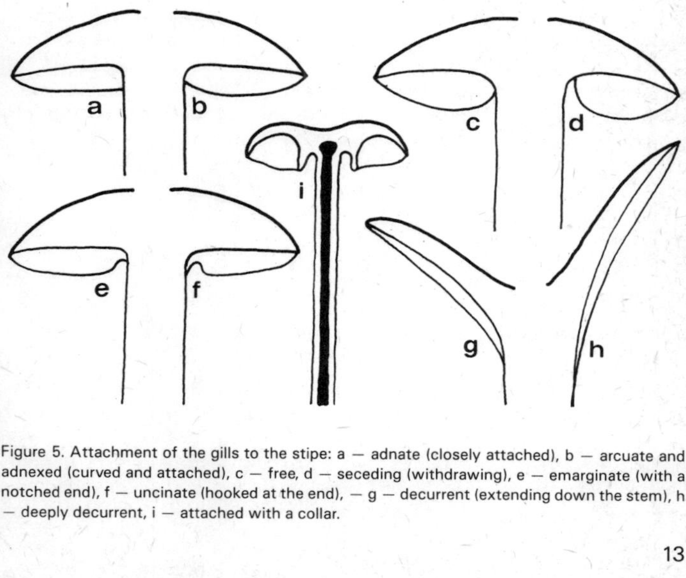

Figure 5. Attachment of the gills to the stipe: a — adnate (closely attached), b — arcuate and adnexed (curved and attached), c — free, d — seceding (withdrawing), e — emarginate (with a notched end), f — uncinate (hooked at the end), — g — decurrent (extending down the stem), h — deeply decurrent, i — attached with a collar.

The walls of the hyphae are made of chitin, enclosing protoplasm. Walls separating individual hyphae are perforated, so that the protoplasm of the whole fungus in most cases forms a cohesive whole.

The *hymenium* is the spore-producing layer of Basidiomycetes (see Figure 8). It is composed mainly of *basidia*, the spore-producing organs, on gills, tubes or spines. Basidia consist of one or more cells, usually club-shaped. A *sterigma* (tiny stalk) connects the spore to the basidium. Each basidium bears, as a rule, four spores, which fall off the sterigmata when ripe. Besides the basidia the hymenium contains similar-looking sterile organs called *basidiola.* In some genera the hymenium contains *cystidia,* sterile organs growing either from the hymenium or directly from the flesh. They vary in form, being most often shaped like spindles, clubs, bottles, or cylinders, and they are usually carried above the surface of the hymenium. Their size, shape, and colour are very important for identifying some genera and species of Basidiomycetes.

Under the hymenium, which is about 10-80mm (½-3in) thick, there may be a thin layer called the *subhymenium.* Under that is the flesh of the gills or other parts of the hymenophore. Gill flesh, or *trama*, is differentiated according to the arrangement of the hyphae forming it.

The spore-producing layer of Ascomycetes of the order Pezizales is called the *thecium*. In the cup fungi (Discomycetes) the thecium lies most often inside the *apothecium*, the cup-shaped fruiting body, or it may be on the upper side of lobed or dimpled caps. It consists of *asci* (sacs), cylindrical or club-shaped cells in which the spores are produced. The ascus bursts open to emit the spores. Between the asci are sterile filaments called *paraphyses.*

The spores of Basidiomycetes and Ascomycetes vary in shape, surface, colour, and size (see Figure 9). These characters are very important for identification. Spores consist of several layers of membrane enclosing cytoplasm, a transparent fluid.

Spores and other microscopic organs are measured in micrometres (μm), equalling 0.001mm or one millionth of a metre.

Classification and Nomenclature of Fungi

Fungi, like other living things, are classified according to a definite system. This system is still being developed, and is often changed and improved as a result of latest scientific research. Older systems of

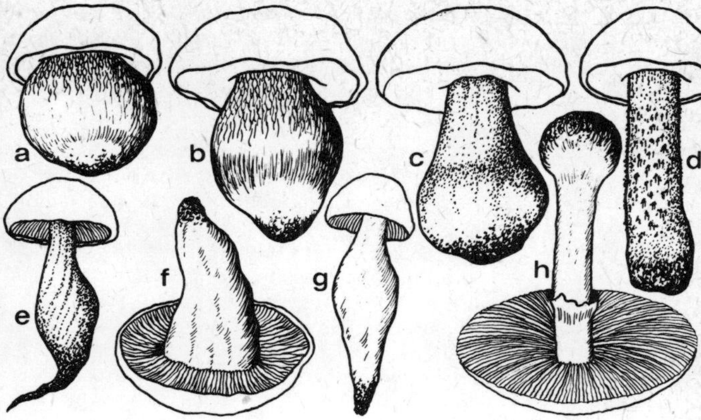

Figure 6. Different shapes of stipe: a — Barrel-shaped, b — ventricose (swollen), c — club-shaped, d — cylindrical, e — spindle-shaped (fusiform) with root-like base, f — conical, g — spindle-shaped, h — cylindrical, with bulbous base.

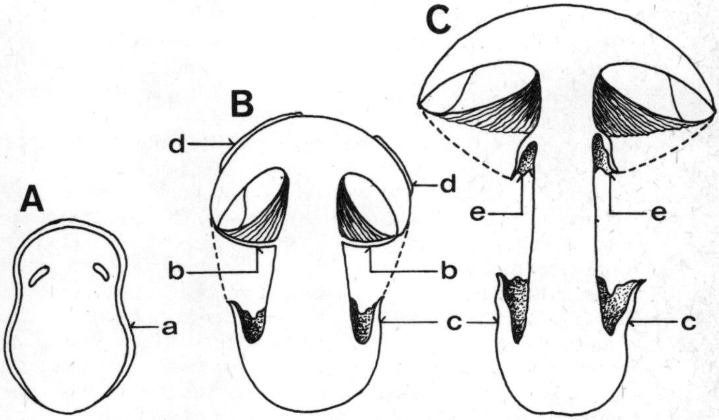

Figure 7. Development of the Death Cap: A — ovate state (the fruiting body is completely covered by universal veil); B — state after tearing of universal veil (its fragments remain on the base of the stipe and form free volva, and with large fragments on the cap; C — fully developed fruiting body with torn partial veil: a — universal veil, b — partial veil, c — remains of universal veil at the base of the stipe (free volva), d — the remains of universal veil on the cap (fragment), e — remaining part of partial veil on the stipe (ring).

classification were quite artificial, and in most cases were based on similarity of appearance of the fruiting bodies. The modern system is based on the similar development of related groups, as it is among animals and higher plants.

The world of fungi is divided into seven classes:

Chytridiomycetes are microscopic, single-celled fungi with many nuclei. Their whirling spores and sexual cells each have one filament at the rear. Most of these fungi live in water, but some are found in soil.

Oomycetes are also microscopic, single-celled fungi. They have a fibrous structure with no divisions, and the whirling spore has a double filament. They live as saprophytes or parasites in fresh water and soil. Some are parasites on species of cultivated plants.

Zygomycetes have well-developed, divisionless hyphae, with spores that do not whirl. Most are saprophytes living in or on the soil. As decomposers they are very important for the creation of humus. A few species are parasites on insects.

Endomycetes are microscopic fungi, some single-celled, some multi-celled. They reproduce by budding. Some are saprophytes in soil, on plant debris, in the digestive organs or on the secretions of animals. Some parasitise higher plants.

These first four classes are of no practical importance to mushroom pickers.

Ascomycetes have well-developed hyphae with partitions, and each hypha cell has a single nucleus. They produce multiple fruiting bodies with asci containing spores. Some are saprophytic on plant or animal remains. Others are parasites on higher plants. This class contains many species, including several that are important for practical mushroom-pickers, for example morels and truffles. Some of them are described in this book.

Basidiomycetes have well-developed hyphae with partitions. Each cell contains two nuclei. The fruiting bodies have a variety of shapes, and in their hymenia are basidia with non-whirling spores. Most of the Basidiomycetes are saprophytes and rarely parasitic. This class contains the majority of the fungi described in this book.

Deuteromycetes have partitioned hyphae developed like those of Ascomycetes or Basidiomycetes. As it has not yet been proved whether they reproduce sexually, it is still not clear whether they belong to the Ascomycetes or the Basidiomycetes. Their classification is not based on their natural affinity and development, but on their morphological similarity (appearance). It is therefore an artificial system. They are not important for mushroom-pickers, but they are interesting from the point of view of their abnormal development.

Each fungus, like an animal or a higher plant, has its own scientific name. This name is binomial, that is, it consists of two words, the genus and the species. The genus, which takes a capital letter, comes first. By convention, they are printed in italics, thus: *Agaricus campestris* (the Field Mushroom).

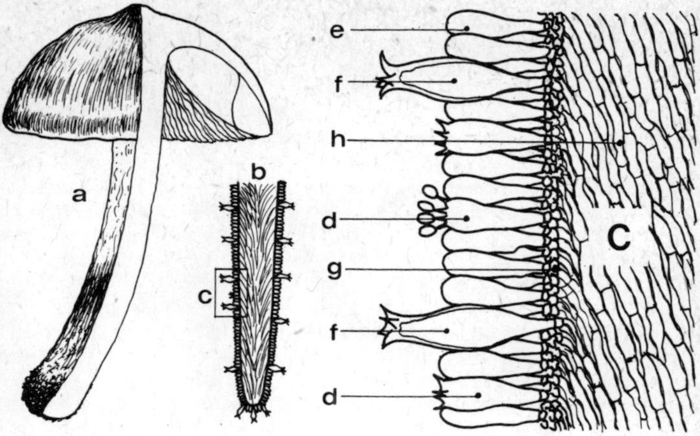

Figure 8. Structure of the hymenium of gill-bearing fungi: a — fruiting body, b — transverse cut through a gill, c — part of the hymenium, d — basidium, e — basidiole, f — cystidium, g — subhymenium, h — trama (flesh) of the gill.

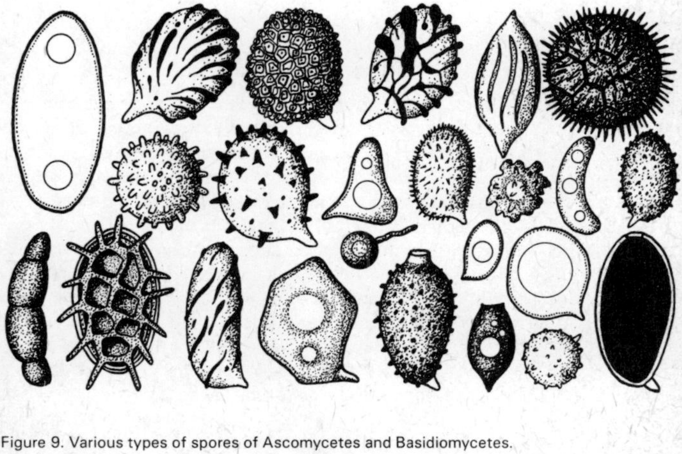

Figure 9. Various types of spores of Ascomycetes and Basidiomycetes.

The scientific names of fungi strictly follow agreed international rules of nomenclature. The scientific name is followed by the name (or an abbreviation of the name) of the person who first described the species. Two works serve as a basis for the scientific names of fungi. Person's *Synopsis methodica fungorum*, written in 1801, covers the Gasteromycetales (a division of the Basidiomycetes), and Elias Fries's *Systema mycologicum* (1821) the rest of the fungi. Names proposed by earlier authors are allowed only if Persoon or Fries mentions them. For example, the High Morel was described for the first time by Fries in 1821, and its correct scientific name is *Morchella elata* Fr. The Grey Boletus was described for the first time by Schaeffer in 1763, and Fries in his *Systema mycoligicum* accepted the name that Schaeffer gave it, so the correct scientific name of this fungus is *Boletus luridus* Schaeff. ex Fr.

If a fungus has been transferred to a different genus, the name of the original author is put in parentheses, and the name of the author who has made the change follows. For example, the Elegant Boletus was first described by Klotzsch in 1832 and given the name *Boletus grevillei*. In 1945 Singer transferred this fungus to the genus *Suillus*, and its correct name is now *Suillus grevillei* (Klotzsch) Sing.

If a species was described by two or more authors independently and by different names, the correct name is that which was published first. All the other names may be used only as synonyms.

How to Pick Mushrooms

There are certain basic rules about mushroom picking, especially in regard to the environment. You should pick only those mushrooms you want to take home. Species you are not interested in should be left as they are, undamaged; maybe somebody else will want them. In woodlands you should move quietly so as not to disturb the wildlife.

Pick the fruiting bodies as they are. Seize them by the stipe, and give them a twist to loosen them, until they separate completely from the substrate. A wicker basket is the most suitable carrier to put the mushrooms in: do not use plastic bags or any other soft materials which might speed up fermentation and deterioration.

If the stipe is brittle and easily broken, loosen the base from the earth by putting a knife blade under the base of the fungus and levering upwards. After the stipe has been removed it leaves a slight depression in the soil. This must be covered with moss or humus, pressed into the cavity to prevent the mycelium from drying up.

Before putting the mushroom in the basket remove any dirt or bits of soil from it, and also peel off the slimy cuticle of the flesh so as to avoid making the other fungi dirty. Put large and fleshy fungi in the bottom of the basket, and place the soft and brittle ones on top so that they are not broken or crushed. Damaged mushrooms do not look so pleasing.

Mushroom-pickers are often unsure whether to pick fungi completely, including the base of the stipe, or to cut off the stipe close to the ground. Both ways are right, but with unknown fungi it is advisable to pick the whole fruit down to the base because they are more easily identified by a specialist when complete.

Fungi must be dealt with immediately you get them home, while they are still fresh. Clean them carefully, cut them up if they are to be dried, prepared them for pickling, or cook them and eat them. Clean and prepared fresh mushrooms can be stored in a refrigerator, but they must be cooked and eaten soon after being taken out of it.

When and where to go Mushroom-picking

It is possible to go mushroom-picking almost all the year round, though the mushroom season lasts from about May to November. There are several good edible fungi that can be found in a mild winter without long periods of frost. These are the so-called winter mushrooms, for example the Oyster Fungus (*Pleurotus ostreatus*), and the Winter Mushroom, also called the Velvet-Stemmed Agaric or Velvet Foot (*Flammulina velutipes*). Both these fungi are woodland mushrooms, and they usually grow in large clumps on the trunks of deciduous trees.

In the main fungus season there are typical spring, summer, and autumn fungi. The most common spring mushrooms are the Spring Hygrophorus (*Hygrophorus marzuolus*), various species of Morels, such as *Entoloma clypeatum*, St George's Mushroom (*Calocybe gambosa*), and among poisonous species the False Morel /*Cyromitra esculenta*).

The most important summer mushrooms are various species of boletus, for example, the Summer Boletus (*Boletus aestivalis*), the Royal Boletus (*Boletus regius*), and *Boletus appendiculatus*; the most tasty species of Russula, such as the Bare-Edged Russula (*Russula vesca*), the Blue and Yellow Russula (*Russula cyanoxantha*), and the Green Cracking Russula (*Russula virescens*); the Parasol Mushroom (*Macrolepiota procera*), *Amanita caesarea* and the Downy Boletus (*Xerocomus subtomentosus*). The main poisonous summer fungi are the Death Cap

(*Amanita phalloides*), the Panther, called also False Blusher or Panther Cap (*Amanita pantherina*), the Livid Entoloma (*Entoloma* [Rhodophyllus] *sinuatum*) and the Red-staining inocybe (*Inocybe patouillardii*).

The richest crop of mushrooms grows in forest in autumn. Among the most tasty ones are the Penny Bun or Cep (*Boletus edulis*), *Boletus pinophilus*; the Wood Blewits (*Lepista nuda*), the Blewits (*Lepista saeva*), the Firwood Agaric, called also Man on Horseback (*Tricholoma flavovirens*), the Dingy Agaric (*Tricholoma portentosum*), the Saffron Milk Cap (*Lactarius deliciosus*), the Orange Cap Boletus (*Leccinum aurantiacum*), the Cow Fungus, called also Brown Birch Boletus (*Leccinum scabrum*), the Bay Boletus (*Xerocomus badius*), various species of champignons (*Agaricus*), and species of *Suillus* and *Gomphidia*. Poisonous autumn mushrooms are especially the Panther (*Amanita pantherina*), the Fly Agaric (*Amanita muscaria*), *Clitocybe dealbata, Clitocybe cerussata,* the Orange-brown Corinarius (*Cortinarius orellanus*) and the Brown Roll-rim (*Paxillus involutus*).

There are certain species of fungi that grow continually from spring till late autumn. The outstanding edible species are the Fairy Ring Mushroom or Clover Windling (*Marasmius oreades*), Chantarelle (*Cantharellus cibarius*), the Changeable Agaric (*Kuehneromyces mutabilis*) and the Shaggy Cap (*Coprinus comatus*).

Many factors influence the growth of fungi. The most important are frequency of fruiting, sufficient moisture and suitable temperature. If the changing seasons follow a normal pattern and there are adequate climatic conditions, many fungi appear between the second half of June and the first half of July and then from September until the end of November. Most species grow in the greatest profusion in autumn, but in exceptional climatic conditions their growth can be considerably held back.

The occurrence of certain species is closely dependent on various partnerships with higher plants, as it is on the geological substrate. In mountainous coniferous woods you may often find completely different fungi from those that can be found in lowland deciduous woods. In the same way there is a difference between the fungi growing in mountainous pastures and those growing in lowland meadows. For some species limey soil is suitable, for others an acid soil. But there are species that can grow everywhere because they tolerate the chemical composition of a wide range of soils.

In mountainous coniferous woods look for the Penny Bun or Cep (*Boletus edulis*), Saffron Milk Cap (*Lactarius deliciosus*), Chantarelle (*Cantharellus cibarius*), Larch Boletus (*Suillus grevillei*), *Boletus erythropus*, *Russula olivacea* and Bay Boletus (*Xerocomus badius*). The poisonous fungi include the Destroying Angel or Fool's Mushroom (*Amanita virosa*), the Fly Agaric (*Amanita muscaria*), *Amanita regalis* and *Amanita pantherina* var. *abietinum*.

On the lower slopes and in deciduous woods the edible mushrooms include the Honey Fungus (*Armillariella mellea*), the Gypsy (*Rozites *caperata*), *Hydnum imbricatum*, Wood Hedgehog (*Dentinum repandum*), *Leccinum nigrescens, Boletus pinophilus,* Horse Mushroom (*Agaricus arvensis*), the Miller, also called Plum Agaric or Sweet-bread Mushroom (*Clitopilus prunulus*), and many others.

In lowland oak woods there are plenty of Summer Boletus (*Boletus aestivalis*), the Lurid Boletus (*Boletus luridus*), *Leccinum quercinum*; the Downy Boletus (*Boletus subtomentosus*), the Chantarelle (*Cantharellus cibarius*) and various tasty species of Russula.

In lowland pine woods the main fungi are the Firwood Agaric or Man on Horseback (*Tricholoma flavovirenes*), the Dingy Agaric (*Tricholoma portentosum*), the Bay Boletus (*Xerocomus badius*), *Lactarius deliciosus* var. *Pini*, the Parasol Mushroom (*Macrolepiota procera*), *Gomphidius rutilus*; the Brown Wood Mushroom (*Agaricus sylvaticus*) and the Blusher (*Amanita rubescens*).

In damp open woods look for Morels, the White Popular Boletus (*Leccinum duriusculum*), the Orange Cap Boletus (*Leccinum aurantiacum*) and the Shaggy Cap (*Coprinus comatus*).

In deciduous woods in warmer regions the edible species include the Royal Boletus (*Boletus regius*), the Bronze Boletus (*Boletus aureus*), and *Amanita caesarea*, and also such poisonous species as the Devil's Boletus (*Boletus satanas*), Purple Boletus (*Boletus rhodoxanthus*) and the Death Cap (*Amanita phalloides*).

In meadows you will find St George's Mushroom (*Calocybe gambosa*), Blewits (*Lepista saeva*), *Lepista irina*, the Pink Bottom or Field Mushroom (*Agaricus campestris*), the Fairy Ring Mushroom or Clover Windling (*Marasmius oreades*) and *Calvatia utriformis.*

In gardens and parks there grow various species of *Agaricus,* and *Lycophyllum decastes*, the Oyster Mushroom (*Pleurotus ostreastus*), *Entoloma clypeatum* and the Fairy Ring Mushroom (*Marasmius oreades*).

This is only a very small sample of the fungi that grow in specific environments. It is a great pity that most of this wealth of fungi remains unused.

Unfortunately it is not possible to give general advice when and where to go mushroom-picking so that you return with a rich haul. Each mushroom-gatherer knows certain places to return to year after year and exactly when and what species of fungi to look for. People like mushroom-picking in those places because they are sure about the mushrooms growing there, and sure of bringing home the mushrooms they like. When they come to unknown regions and have no one to give good advice about the mushrooms growing there, they must rely on luck. If you know many species of good edible fungi you do not have to worry that you will go home with an empty basket, and you may be sure that you will always find some specimens that are best for eating.

BEWARE OF POISONOUS FUNGI

The unhappy outcome of uninformed mushroom-picking is poisoning. In many cases such poisoning is due to ignorance or a superficial approach to identification. In comparison with the edible or harmless species there are not many poisonous ones. Therefore everyone who wants to pick fungi should learn to distinguish at least 20-25 dangerous species.

In order to preserve the pleasure of mushroom-picking and to avoid the danger of poisoning, pick only those species which you know for certain. It is very dangerous to pick even one species that you know superficially only, and it can be fatal.

There is no straightforward rule to distinguish an edible fungus from a poisonous one. In the past people believed that silver cutlery or onion turned black in contact with a poisonous fungus. This is not true. People also thought that all poisonous fungi stink and taste acrid or bitter. The fact is that the most dangerous fungi such as the Death Cap (*Amanita phalloides*), the Panther (*Amanita pantherina*) and the Livid Entoloma (*Entoloma sinuatum*) have a mild flavour and no smell. It is also untrue that the species which snails and insect larvae eat and live on can also be eaten by Man. The fact is that you may often find the Death Cap (*Amanita phalloides*) perforated by larvae. The most effective protection against poisonous fungi is recognizing them accurately and not picking them.

The most dangerous fungi are the Death Cap (*Amanita phalloides*) and the two species akin to it, *Amanita verna* and the Destroying Angel (*Amanita virosa*). These three toadstools cause the majority of fatal poisonings.

Poisoning by the Death Cap is very serious because the first symptoms are not felt until 8-12 hours after eating it. At first the patient feels sick, dizzy, and lethargic, has a headache, and breaks into a cold sweat. Then acute vomiting and severe diarrhoea follow, which may last 2 to 3 days and exhaust the patient. After several hours' apparent remission sharp abdominal pains are followed by jaundice. Soon the patient falls into coma and dies. In some cases of slight poisoning the patient recovers, but only very slowly, and the consequences last a very long time, often all his life.

Poisoning by the Panther (*Amanita pantherina*) is frequent but for the most part not fatal. Approximately 10 to 15 per cent of all cases result in death. If a person poisoned by the Panther gets prompt medical help recovery is usually successful.

Similar, but in many cases milder, poisoning is caused by Fly Agaric

(*Amanita muscaria*) and *Amanita regalis*. However, poisoning by Fly Agaric is very rare because almost everybody knows it well and avoids it.

Slightly poisonous fungi are *Amanita gemmata* and *Amanita porphyria*.

The Livid Entoloma (*Entoloma sinuatum*) is highly poisonous. Symptoms come on at any time after 20 minutes to 4 hours at the latest. The patient vomits, has diarrhoea, stomachache and headache, feels thirsty and very weak. In cases of slight poisoning these symptoms slowly disappear and are gone completely after several days. In higher concentration Livid entoma may also cause death.

Besides the Livid Entoloma there are other poisonous species of this genus, for example *Entoloma rhodopolium* and *Entoloma nidorosum*.

The *Cortinarius orellanus* is particularly dangerous. The first symptoms of poisoning by this toadstool appear usually very late, after 3 to 14 days. Its poisonous components damage first of all the kidneys and the liver. At the beginning the poisoned person feels very thirsty, with a dry and burning mouth. Later nausea, vomiting, headache and abdominal pains with constipation and fever set in. The symptoms become worse and death comes in 2 to 3 weeks, or sometimes after several months.

Among the Icocybes there is a high proportion of poisonous species, for example *Inocybe fastigiata* and *Inocybe geophylla*. The most poisonous is the Red-staining Inocybe (*Inocybe patouillardii*). The symptoms of poisoning by this fungus are described on page 118.

Poisoning similar to that of *Inocybe* is caused by some species of white *Clitocybe, Clitocybe dealbata, C. rivulosa* and *C. cerussata*. The symptoms of poisoning appear usually in 15 to 20 minutes. The affected person at first feels sick, goes into a cold sweat, suffers from fever, vomiting and diarrhoea. At the same time the patient suffers from breathing difficulties, heart trouble, dizziness and sight defects, all without losing consciousness. In severe poisoning, death may occur in a few minutes; the patient dies from suffocation or heart failure.

Of the *Tricholoma* genus the most poisonous is Tiger Tricholoma (*Tricholoma pardalotum*) which rapidly causes painful symptoms; within an hour the patient develops cramps, vomiting and diarrhoea, abdominal pains and headache. These symptoms last from 2 to 6 hours. Though the poison leaves no lasting effects it is very unpleasant. Other poisonous *Tricholoma* species include *Tricholoma pessundatum, T. sejunctum* and *T. sulphureum*.

Of the few poisonous fungi growing on wood the best known are *Omphalotus olearius* and the Sulphur Tuft (*Hypholoma fasciculare*). While poisoning by *Omphalotus olearius* is quite mild, poisoning by Sulphur Tuft may have serious consequences.

Dangerous poisoning may be caused by the False Morel (*Gyromitra esculenta*). It is a very insidious fungus; there are many people who eat it for years without any harm, while it makes other people seriously, even fatally ill, especially children. The toxic component of this fungus, called gyromitrine, has still not been fully studied. Some scientists think that it is created in old fruiting bodies, probably through the destructive activity of micro-organisms. Scalding the mushrooms before cooking them, as used to be recommended, is not effective.

Some years ago the Brown Roll-rim (*Paxillus involutus*) was added to the index of poisonous fungi. In the literature it was usually described as an edible fungus. In 1963, after several cases of poisoning, it was proved by scientists that frequent consumption of Brown Roll-rim causes destruction of the red blood-corpuscles.

The Devil's Boletus (*Boletus satanas*) is poisonous only when consumed raw or partly cooked. To the inexperienced its name makes them think it must be the most dangerous species. The fact is that although poisoning by the Devil's Boletus is very unpleasant (vomiting, diarrhoea, spasms), it does not cause death and leaves no after-effects. Similar effects are caused by the Purple Boletus (*Boletus rhodoxanthus*), a fungus akin to the Devil's Boletus.

Coprinus atramentarius causes poisoning as well, but only when eaten with any sort of alcoholic drink. Otherwise it is harmless.

The Common Earthball (*Scleroderma citrinum*) is poisonous only when consumed in great quantity. Two or three thin slices of it are harmless.

According to the latest researches and experience *Clitocybe inversa, Mycena pura* and *Collybia futsipes* are slightly poisonous.

The same is true of *Agaricus xanthodermus* and *Agaricus placomyces.*

Also mildly poisonous are acrid-tasting species of Russula, which include *Russula sardonia, Russula queletii, Russula emetica, Russula foetens, Russula atropurpurea* among many others.

Lactarius turpis and *Lactarius helvus* may cause slight stomach upset.

Even mushrooms generally considered safe may cause trouble if they are gathered when damaged or past their prime. Young mushrooms may deteriorate, producing toxins if they are not cooked promptly. Spoilt edible mushrooms contain nonspecific poisons formed by autolysis (cell destruction by enzymes) or the destructive activity of bacteria.

When any symptoms of mushroom poisoning appear you must call a doctor at once. Before he arrives do your best to get the meal that caused poisoning out of the system. It is advisable to drink lots of warm water to provoke vomiting. After having vomited it is necessary to use a purgative, such as castor oil. The poisoned person must drink plenty of liquid so as to dilute the poison in the gut. The best treatment is washing out the stomach, which should be done by a doctor.

MUSHROOMS IN THE DIET

In ancient Rome meals which included mushrooms were highly appreciated. Ancient literature, natural sciences, medical literature and fiction are all proof of this fact. Mushroom dishes are becoming increasingly popular these days as scientific knowledge about fungi offers everyone a new look at this food source.

Until recently opinions on the nutritious content of mushrooms were often contradictory. Some people underrated them as having no food value; others considered them to be the equivalent of animal protein. According to the results of the latest scientific researches, the nutritious value of mushrooms lies somewhere in the middle of these two extreme views.

Even though the calorie value of mushrooms is low by contemporary standards, it should not be neglected. When cooked with care they are a valuable supplement to the diet. Besides their delicious taste and exciting smell they contain some important nutrients.

The proportion of digestible protein of most edible mushrooms is low, varying from 3.5 to 25 per cent. However, they contain some essential amino-acids. The most important vitamins in them are B1, B2 and D. According to the latest researches there is also a small amount of vitamin C. They are particularly rich in phosphorus and potassium.

Mushrooms can enrich the diet and relieve the monotony of a menu. They act favourably on the digestion, and when properly cooked they are a useful addition. They should not be served to persons with kidney or stomach trouble or to anyone allergic to them. They are successfully used in slimming because they produce a feeling of fullness and they keep the digestive system busy.

Prepare only fresh, sound mushrooms and cook them carefully. Mushrooms with tough flesh should be cut into very small pieces and cooked longer than the soft and brittle species. Mushrooms are harmless when reheated provided they were put in a refrigerator soon after they were cooked. However, we do not recommend you to keep food containing mushrooms; they are best eaten soon after being cooked. Serve them to children under ten years old only in small amounts.

Fresh mushrooms are only a seasonal food, but they can be preserved for winter use. The most common methods are drying and pickling them or keeping them in salt.

For drying, choose sound specimens. Do not wash them but scrape clean with a knife. The cleaned mushrooms are cut into thin slices parallel with the longitudinal axis of each mushroom. Do not separate cap from stipe. Lay the cut slices, not touching each other, on a stainless wire mesh. Dry the mushrooms in a well-ventilated room, in shade or in the sun. In the country, people dry them in ovens. When mushrooms are dried in the open air make sure they are protected from dew and rain. They are dry when the slices are easily crumbled between the fingers. The most suitable for drying are the various species of tube-bearing fungi, because their flesh is usually tough. They should be stored carefully when dried to prevent them from becoming mouldy or nibbled by insects. The best storage vessels are glass jars or tins with close-fitting lids.

Another way of preserving mushrooms is to pickle them in boiling vinegar with or without sugar. The most suitable for pickling are the tube-bearing fungi, such as Chanterelles and *Lactarius*. Choose young and undamaged fungi only; clean and wash them carefully. Pickle small ones whole but cut the bigger ones into quarters, leaving the cap attached to the stipe. With Russulas, champignons and *Armillariella* only the caps are pickled as a rule. Before pickling, soak the clean mushrooms in boiling water with a little salt and citric acid for 10-15 minutes, according to their toughness. Allow the cooked mushrooms to drip dry and then put them into clean jars. Pour over the vinegar so that the mushrooms are covered. On top you can put a bay leaf, a slice of horseradish, or a slice of onion, but only to mushrooms that do not have a slimy cuticle. The liquid is filled up to 2cm (1in) from the top of the jar. For sterilisation use Kilner jars or other suitable vessels. Lay a piece of wood or a tile in the bottom of a large saucepan, and stand the filled and sealed jars on it. Fill the pan to the top, and sterilise the mushrooms for about 30 minutes at 100°C (212°F). Take the jars out when partly cooled, check that they are still sealed and leave them to cool completely. Any jars that are not sealed will have to be sterilised again.

Recipes for pickling solutions

Spiced vinegar

For 1-1.5kg (2¼-3½lb) of mushrooms you will need:

Metric	*Imperial*	*American*
0.5 litre water	18 fl oz water	2¼ cups water
0.5 litre vinegar	18 fl oz vinegar	2¼ cups vinegar
½ tsp salt	½ tsp salt	½ tsp salt
20 black peppercorns	20 black peppercorns	20 black peppercorns
Mustard seed to taste	Mustard seed to taste	Mustard seed to taste

Sweet spiced vinegar

⅓ litre water	12 fl oz water	1½ cups water
1/6 litre vinegar	6 fl oz vinegar	¾ cup vinegar
1 onion	1 onion	1 onion
2 cubes of sugar	2 tsp sugar	2 tsp sugar
10 black peppercorns	10 black peppercorns	10 black peppercorns
5 white peppercorns	5 white peppercorns	5 white peppercorns

A bayleaf, ginger, mustard seed and salt to taste.

If you add a higher proportion of onion the liquid thickens and tends to jellify.

Mushrooms can also be preserved in salt. In this way they keep their original taste and smell, but they are best used only in soups. Select only young, firm mushrooms and clean them with a damp cloth. Cut them up and leave them to dry a little. Then mix the slices with salt and pack them firmly into jars. On top put a final layer of salt, and cover with cellophane or clingfilm. For 1kg (2¼lb) mushrooms you will need 100-150g (3½-5oz) of salt. Most mushrooms can also be preserved in brine. Pack the prepared mushrooms into jars and pour on a 50 per cent brine to cover them. They must be well rinsed in running water before use and they will need no salt in cooking.

Mushroom Recipes

Mushroom Open Sandwiches

Metric/Imperial	*American*
150g/6oz fresh mixed mushrooms	6oz fresh mixed mushrooms
50g/2oz pickled mushrooms	2oz pickled mushrooms
30g/1¼ oz potato flour	⅓ cup potato flour
50g/2fl oz milk	¼ cup milk
100g/4oz butter	1 stick of butter
Fresh parsley	Fresh parsley
50g/2oz Emmenthal cheese	⅓ cup Emmenthal cheese
100g/4oz salami	4oz salami
white bread, salt	white bread, salt

Cut the fresh mushrooms into slices and stew with a little salt. Mix the milk and potato flour and add to the mushrooms to thicken the juice. When cool spread on slices of bread and butter; decorate with pickled mushrooms, cheese, salami and parsley.

Spring Open Sandwiches with Morels

250g/9oz various species morels	9oz morels
40g/1½ oz butter	3 tbsp butter
50g/2oz grated cheese	⅓ cup grated cheese
2 hard-boiled eggs	2 hard-cooked eggs

White bread, fresh parsley, beetroot juice, black pepper, coriander, clove, salt

Fry slices of one-day-old bread in butter. Fry the mushrooms, finely chopped, with onion, pepper, and coriander; spread them on the fried bread and top with a mixture of grated cheese, finely-chopped egg yolk, parsley, and the egg whites coloured with beetroot juice. You can add a slice of beetroot or green peppers. These open sandwiches are nicest served with a dish of stewed morels or with other mushrooms preserved in oil or vinegar.

Mushrooms à la Provençale

(Use St George's Mushroom, Dingy Agaric or Saffron Milk Cap)

Metric/Imperial	*American*
40g/1½ fl oz vegetable oil	3 tbsp vegetable oil
60ml/2 fl oz white wine	¼ cup white wine
150ml/¼ pint beef stock	⅔ cup beef stock
150g/5oz sliced ham	5oz sliced ham
4 slices of bread	4 slices of bread
½ lemon, salt	½ lemon, salt

Cut the mushrooms into thin strips and braise them in half the oil. Gradually stir in the beef stock and the white wine. Place the mushrooms on slices of hot baked or grilled ham, and serve with bread fried in the rest of the oil and flavoured with lemon juice.

Parasol Mushrooms with Garlic

(Instead of Parasol Mushrooms you can use the Blusher or Summer Boletus)

200g/8oz fresh mushrooms	8oz fresh mushrooms
50g/2oz butter or oil	½ stick butter or ¼ cup oil
4 cloves of garlic	4 cloves of garlic
salt	salt

Slice the mushrooms into large pieces, spread them with garlic crushed with salt and grill them in a pan with the butter or oil until crisp. Serve with a tomato and cucumber salad.

Stuffed Mushrooms

250g/9oz large mushrooms	9oz large mushrooms
100g/4oz ham or smoked tongue	4oz ham or smoked tongue
50g/2oz butter	½ stick butter
1-2 small onions	1-2 small onions
125ml/¼ pint milk	⅔ cup milk
50g/2oz flour	½ cup flour
50g/2oz Parmesan cheese	⅓ cup Parmesan cheese
20g/1oz fresh breadcrumbs	½ cup fresh breadcrumbs
250g/8oz rice	1 cup rice
salt,	salt
lemon juice	lemon juice

Wash and peel the mushrooms and remove the stalks. Scrape out the inside of the caps and sprinkle them with lemon juice; fry quickly in butter until brown. Mix the scraped out part with butter and finely cut onion and fry until golden-brown; add finely cut ham or tongue and the finely chopped stalks of the mushrooms.

Prepare a white sauce with butter, milk and flour and stir the mushroom mixture into it. Pile the whole mixture on to the caps, sprinkle them with salt, grated Parmesan cheese and breadcrumbs, pour over a little melted butter, and bake them in a hot oven for a few minutes. Serve with boiled rice, decorated with slices of stewed mushroom and fresh parsley.

Mushrooms Rolled in Ham

(Use Firwood Agaric, Dingy Agaric, Honey Fungus and species of *Russula*)

Metric/Imperial	*American*
300g/12oz mushrooms	12oz mushrooms
50g/2oz bacon fat	2oz bacon fat
200g/8oz ham or salami	8oz ham or salami
2 onions	2 onions
black pepper, salt	black pepper, salt

Chop the mushrooms finely and cook them gently with the chopped onion in bacon fat. Add salt and black pepper to taste. When the mushrooms are soft, put them aside to cool. Spoon the mushroom, bacon and onion mixture on to slices of ham or salami, and make them into rolls. Put the rolls on a greased baking tray and bake until done. Serve with bread.

Borsch with Mushroom Blintzes

The Borsch

200g/8oz mushrooms, assorted	8ox mushrooms, assorted
250g/12oz pork or beef bones	12oz pork or beef bones
1 very small onion	1 very small onion
30g/1oz butter	2 tbsp butter
20g/½oz flour	2 tbsp flour
75g/3oz mixed, diced vegetables	¾ cup mixed, diced vegetables
500g/1¼lb beetroot	1¼lb beets

Vinegar, bayleaf, black pepper, sugar, salt

Wash the beetroot, boil in the skin and peel when cool. Make a stock by boiling together the bones, vegetables (to your own choice), onion and sliced mushrooms, with the bayleaf and a little pepper. Strain the stock, add vinegar to taste. Melt the butter, add flour and cook till light brown; add the strained stock, vegetables, onion, mushrooms and beetroot. Season as required with salt, pepper and sugar.

The Blintzes
Pastry

Metric/Imperial	*American*
120g/4½oz flour	1 cup flour
1 egg yolk	1 egg yolk

Water and salt

Make a soft dough and leave it covered for a while. Roll it out thinly and cut into small squares. Fill the squares with the stuffing (below), and fold them into triangles, pinching the edges together. Boil the blintzes in salt water for about 10 minutes.

The Stuffing

400g/14oz mushrooms (*Boletus* or *Suillus* species)	14oz mushrooms
30g/1oz onion	1oz onion
20g/¾oz butter	1½ tbsp butter
1 eggwhite	1 eggwhite
1 tbsp breadcrumbs	1 tbsp breadcrumbs

black pepper, salt

Boil the mushrooms and chop them up finely, then braise them with chopped onions in butter. Add breadcrumbs, eggwhite, salt and pepper.

Serve the blintzes with the broth.

Bean Soup with Mushrooms and Cream

200g/8oz mushrooms (*Russula* and *Boletus*)	8oz mushrooms
250g/10oz beef bones	10oz beef bones
1 medium onion	1 medium onion
100g/4oz haricot beans	1 cup haricot beans
50g/2oz mixed vegetables	½ cup mixed vegetables
125ml/¼ pint cream	⅔ cup cream
salt	salt

Boil the bones to make stock, add the onion and mixed diced vegetables before cooking is finished. Soak the beans in water overnight, boil until soft, drain, and add the strained stock to them. Add sliced mushrooms and the vegetables from the stock; cook thoroughly. Thicken with cream just before serving.

Egg and Mushroom Broth

Metric/Imperial	*American*
150g/6oz mushrooms	6oz mushrooms
2 eggs	2 eggs
40g/1½oz butter	3 tbsp butter
40g/1½oz flour	⅓ cup flour
water, salt, chives	water, salt, chives

Melt the butter and stir in the flour to make a roux, then add the eggs, well beaten, stir and add cold water. Stew the mushrooms, chop them up and add them to the broth; simmer for 25 minutes. Season and add finely chopped chives.

Cabbage Soup with Mushrooms

180g/8oz mushrooms	8oz mushrooms
280g/12oz cabbage	12oz cabbage
250g/10oz beef bones	10oz beef bones
50g/2oz butter	½ stick butter
20g/¾oz flour	3 tbsp flour
30g/1oz rice	2 tbsp rice
125ml/4fl oz milk	½ cup milk
a small onion	a small onion
50g/2oz green peas or beans	½ cup green peas or beans
chives, salt	chives, salt

Clean and wash the mushrooms, cut them into small pieces, and cook them in some of the butter with chopped onion. Melt the rest of the butter in another pan; add the shredded cabbage and cook it slowly. Boil the bones to make stock, strain it, and thicken with flour combined with the milk. Parboil the rice. Add mushrooms, cabbage, rice, peas or beans and finely chopped chives to the stock, season and cook for a few minutes.

Mushroom Pancakes

(For best results use Oyster Mushroom, *Suillus* species and Blewits)

Metric/Imperial	*American*
200g/8oz mushrooms	8oz mushrooms
500ml/18fl oz milk	2¼ cups milk
3 large eggs	3 large eggs
140g/5oz flour	1¼ cups flour
100g/4oz butter or bacon fat	1 stick butter or ½ cup bacon fat
100g/4oz chicken liver	4oz chicken liver
1 medium onion	1 medium onion

Black pepper, fresh parsley, salt

Chop up the liver and the onion, and fry lightly until the liver is browned and the onion tender. Parboil the mushrooms, chop them finely and mix them with the onion and liver. Season to taste. Keep this mixture warm while you make the pancakes. For the pancakes, separate the eggs and beat the yolks with half the milk; gradually add the flour and mix to a smooth paste; add the remaining milk. Whip the whites until stiff and fold them into the mixture. Make thin pancakes cooked on both sides. Put a spoonful of the mushroom mixture on each pancake and roll it up.

Rice Croquettes with Mushrooms

(Use species of *Boletus, Russula,* and *Tricholoma*)

300g/12oz long-grained rice	1½ cups long-grained rice
250ml/½ pint water	1¼ cups water
100g/4oz mushrooms	4oz mushrooms
1 small onion	1 small onion
40g/1½oz butter	3 tbsp butter
20g/¾oz flour	3 tbsp flour
125ml/4fl oz milk	½ cup milk
100g/4oz dried breadcrumbs	1 cup dried breadcrumbs
150g/6oz fat for frying	¾ cup fat for frying
salt	salt

Put the rice and water in a lidded vessel; cook in the oven for about 20 minutes. Make a white sauce with flour, butter and milk; mix it with the rice. Parboil the mushrooms with chopped onion and add them to the rice mixture. Scoop out spoonfuls of the mixture and form them into rolls about 100mm (4in) long. Roll them in the breadcrumbs and fry in the hot fat until golden brown. Serve the croquettes with fried potatoes and a green, cucumber or tomato salad.

Mushrooms with Cream Cheese

(For the mushrooms use *Boletus* species, Field Mushrooms and St George's Mushroom; any cream cheese will do, but cream cheese made from ewe's milk — failing that, goat's milk — is best).

Metric/Imperial	*American*
500g/1lb mushrooms	1lb mushrooms
50g/2oz butter	½ stick butter
250g/8oz cream cheese	8oz cream cheese
2 medium eggs	2 medium eggs
black pepper or paprika	black pepper or paprika
1 small sweet capsicum	1 small sweet capsicum
salt	salt
a few capers	a few capers

Clean, wash and chop the mushrooms, sprinkle them with salt, black pepper or paprika and chopped capers, and braise in butter until soft. Add the cheese and the beaten eggs, cook until the eggs are done. Garnish with slices of capsicum. Serve with new potatoes.

Mushrooms à la mode Polonaise

80g/3oz dried mixed mushrooms	3oz dried mixed mushrooms
50g/2oz butter	½ stick butter
1 large onion	1 large onion
300g/12oz potatoes	12oz potatoes
250ml/½ pint cream	1¼ cups cream
2 egg yolks	2 egg yolks

Breadcrumbs, black pepper, salt

Soak the dried mushrooms until soft, boil in salted water, strain and slice them. Boil the potatoes and dice them. Chop the onion and fry it in some of the butter; add the mushrooms and continue frying until they brown. Stir in the cream, egg-yolks and seasoning off the heat. Pour in a little of the water in which the mushrooms were boiled. Add the potatoes to the mushroom mixture. Mix well then put it into a greased ovenproof dish, sprinkle with breadcrumbs, and dot the remaining butter on top. Bake in a hot oven until golden brown.

Champignons comme escalopes à la parisienne

(Use *Boletus* species, Field Mushrooms, Parasol Mushroom, *Tricholoma* species, *Lycoperdon perlatum*)

Metric/Imperial	*American*
500g/1lb mushroom-caps	1lb mushroom-caps
500ml/¾ pint milk	1 pint milk
1 medium egg	1 medium egg
80g/3oz flour	¾ cup flour
200g/8oz fat for frying	1 cup fat for frying
salt	salt

Add egg, flour and salt to the milk, and blend to make a fairly thin batter, neither too thick nor too liquid. Sprinkle the mushroom caps with salt, dip them in the batter, and fry them in hot fat until they are golden brown. Serve with potatoes and spinach or peas.

Paprika Mushrooms

(Use *Suillus* species, Parasol Mushroom, Chanterelle)

500g/1lb mushrooms	1lb mushrooms
40g/1½oz butter	3 tbsp butter
30g/1 small onion	1 small onion
1 tsp paprika	1 tsp paprika
120ml/4fl oz sour cream	½ cup sour cream
20g/¾oz flour	3 tbsp flour
salt	salt

Fry the chopped onion lightly in butter, add the salt, paprika and chopped mushrooms. Cook for 10-15 minutes. Blend the flour into the cream and add to the mushrooms: cook until the mixture thickens. Serve with rich, new potatoes, or boiled dumplings.

Mushroom Goulash

(Use species of *Boletus* and *Russula*)

500g/1lb mushrooms	1lb mushrooms
40g/1½oz piece pork or bacon fat	1½oz piece pork or bacon fat
10g/1 tsp lard	1 tsp lard
1 small onion	1 small onion
20g/¾oz flour	3 tbsp flour
clove of garlic	clove of garlic

paprika, caraway seed, black pepper, coriander, salt

Fry the finely chopped onion with the fat until golden, sprinkle in some paprika; wash and chop up the mushrooms and add them to the pan with the crushed garlic. Cook until the mushrooms are soft. Blend in the flour then add stock or water if necessary; add seasoning. Melt the lard, blend in the rest of the paprika, and add to the pan to give a good red colour. Serve with potatoes, pasta or rice.

Mushroom and Vegetable Stew with Dumplings
(Use Field Mushrooms, and species of *Boletus* and *Suillus*)

Metric/Imperial	*American*
500g/1lb mushrooms	1lb mushrooms
100g/4oz green peas	1 cup green peas
200g/8oz cauliflower	8oz cauliflower
50g/2oz piece pork or bacon fat	2oz piece pork or bacon
40g/1½ oz flour	⅓ cup flour
250ml/8fl oz beef stock	1 cup beef stock
30g/1oz butter	2 tbsp butter
20g/1oz fresh breadcrumbs	⅔ cup breadcrumbs
1 medium egg	1 medium egg

Seasonings: fresh parsley, black pepper, ginger, coriander, mace, paprika, salt

Wash the mushrooms, slice the larger ones and cut small ones in half longitudinally; stew them in their own juice for 10 minutes with green peas and the broken-up cauliflower. Blend the flour in melted pork or bacon fat, add the stock, mushrooms, vegetables and seasoning; cook till done. For the dumplings, melt the butter and mix into it the breadcrumbs, beaten egg and chopped parsley. Form into balls the size of a walnut and boil for 5-10 minutes in salted water. Serve stew with the dumplings on top.

Sauce chasseur à la mode polonaise
(Use dried *Boletus* and Bay Boletus)

20g/¾ oz dried mushrooms	¾ oz dried mushrooms
1 medium onion	1 medium onion
50g/2oz dripping	¼ cup dripping
30g/1oz flour	¼ cup flour
100g/4oz salami or garlic sausage	4oz salami or frankfurter
50g/2oz pickled gherkins	2oz pickled gherkins
10g/1 tsp mustard	1 tsp mustard
500ml/1 pint beef stock	2½ cups beef stock

paprika, mace, juniper berries, salt

Cut up the mushrooms and stew them well in a little water. Fry the onion lightly in the dripping and stir into the flour; add the mushrooms, the liquor in which they were cooked and the stock, plus chopped gherkins, chopped meat, paprika, mace and crushed juniper berries. Cook for 10-15 minutes. Add mustard and salt to taste when cooking is nearly completed.

Vegetable Stew with Mushrooms

Metric/Imperial	*American*
300g/12oz mushrooms (*Boletus* spp)	12oz mushrooms
300g/12oz tomatoes	12oz tomatoes
300g/12oz green peppers	12oz green peppers
1 medium onion	1 medium onion
40g/1½oz dripping	3 tbsp dripping
black pepper, salt	black pepper, salt

Fry the chopped onion lightly in the dripping, then add the chopped peppers. Cook for a few minutes. Stew the sliced mushrooms in a little water and add them to the onions and peppers. Finally add the tomatoes, also cut up. Cook until all is tender, stirring frequently, then add salt and pepper. Serve with potatoes, rice or bread.

Escalopes with Mushrooms

200g/8oz field mushrooms	8oz field mushrooms
1 onion	1 onion
500g/1¼lb lean pork	1¼lb lean pork
400g/1lb flour	4 cups flour
100g/4oz butter	1 stick butter

Black pepper, basil, fresh parsley, other seasoning to taste.

Slice the mushrooms and braise them with the chopped onion in half the butter. Cut thin slices of pork, beat them flat, sprinkle them with salt and pepper, flour on both sides, and fry briskly in the rest of the butter. Sprinkle flour over the stewed mushrooms, add boiling water and seasoning, then bring back to the boil. Pour the mushroom sauce over the escalopes. Serve with new potatoes and cucumber, endive or lettuce salad.

Fish with Mushrooms

Metric/Imperial	*American*
250g/8oz mushrooms	8oz mushrooms
700g/1½lb white fish fillets	1½lb which fish fillets
170g/6oz butter	¾ cup butter
40g/1½oz flour	⅓ cup flour
250ml/8fl oz milk	1 cup milk
40g/1½oz grated cheese	½ cup grated cheese
Fresh parsley, salt, lemon juice	fresh parsley, salt, lemon juice

Clean the fish and cut it into portions. Sprinkle the fillets with lemon juice and salt, and fry them in some of the butter. Clean and cut up the mushrooms. Cook them in a little butter with the chopped parsley. Make a white sauce with about 40g/1½oz of the butter, flour and milk, and stir in the mushrooms. Lay the fish on a shallow greased dish, pour the sauce over it, sprinkle with grated cheese and melted butter, and bake in a hot oven until the cheese is browned.

Pork with Mushrooms and Cheese

(Use Field Mushrooms, *Boletus* and *Tricholoma* species)

300g/12oz mushrooms	12oz mushrooms
300g/12oz pork fillet	12oz sirloin pork roast
50g/2oz butter	½ stick butter
3 eggs	3 eggs
500ml/¾ pint milk	2 cups milk
60g/2½oz grated cheese	¾ cup grated cheese
chives or fresh parsley	chives or fresh parsley
salt	salt

Wash and dry the pork, sprinkle it with salt and put it in a baking tray with a little water. Roast it in the oven, turning it from time to time. When done cut into slices and lay them on a shallow greased dish. Stew the mushrooms until tender. Beat the milk and eggs together. Place the mushrooms on top of the slices of meat, pour over the milk and egg mixture, sprinkle with cheese and chopped parsley or chives, and bake for a few minutes in a hot oven.

Liver and Mushrooms

(Use *Boletus* and Field Mushrooms)

Metric/Imperial	*American*
200g/8oz mushrooms	½ lb mushrooms
400g/1lb calves' liver	1lb calves' liver
60g/3oz bacon rashers	3oz sliced bacon
30g/1oz butter	2 tbsp butter
15g/½ oz flour	1 tbsp flour
100ml/4fl oz Madeira	½ cup Madeira
100ml/4fl oz chicken or veal stock	½ cup chicken or veal stock

black pepper, salt

Prepare the liver, cut it into slices with strips of bacon and a few slices of mushroom. Meanwhile wash and chop up the remaining mushrooms and stew them in their own juice. Sprinkle the flour over the liver, add the mushrooms, stock and wine, and cook for a few minutes, seasoning at the last minute.

Meaning of Symbols

Edible	
Inedible	✕
Poisonous	†
Fatal	☠

COMMON MOREL **Edible**

Morchella esculenta L., ex St-Am

Cap Orange-yellow to brown, greyish-brown or grey, usually egg-shaped, rarely irregularly spherical, hollow inside, 30-70mm (1⅛-2¾in) high, 30-60mm (1⅛-2¼in) wide. Its margin joins on to the stipe as it develops. The surface of the cap is covered with irregular shallow depressions bedded with spore-producing layers. The dimples on the cap are separated from each other by cross walls of different height and irregular pattern.

Stipe Almost cylindrical, usually thicker at the base, hollow inside, 30-90mm (1⅛-3½in) long, 25-35mm (1-1⅜in) thick, with creamy or creamy-yellow surface; when ripe it becomes light brown; velvety or slightly scaly, often with longitudinal ridges.

Flesh Whitish, creamy or orange-yellow, rather thin 1.5-2mm; 1/20-1/16in) and fragile. It has a mild flavour and the smell is slight.

Spore deposit Creamy, almost pale ochre.

Spores 19-22×11-15μm, elliptical; smooth; no vegetal oil drops.

Habitat This popular spring fungus grows from the second half of March until the beginning of May in open woodland with plenty of light, or sometimes in parks and gardens. It can also be found under bushes in meadows. As its cap is usually the colour of dead leaves it can be easily overlooked.

Use A good, edible mushroom, suitable especially for soups. Can be stuffed with ground (minced) seasoned meat and baked.

Remarks Before use it is advisable to scald it in boiling water. This mushroom cannot be confused with any poisonous fungus; it could be taken for other varieties of *morchella*, but all of them are edible.

CONICAL MOREL **Edible**

Morchella conica Pers.

Cap Greyish-brown or dark-brown, conical, hollow inside, 30-90mm (1⅛-3½in) high, 20-60mm (¾-2¼in) wide; its margin is attached to the stipe. The surface of the cap is covered with shallow dimples separated from each other by perpendicular and crosswise ribs. The dimples are bedded with a spore-producing layer. Perpendicular ribs run from the top of the cap to its margin, and the dimples form an almost regular pattern.

Stipe Cylindrical, sometimes narrowed at its base, hollow inside, 30-50mm (1⅛-2in) long, 15-35mm (1½-1⅜in) thick, with longitudinal ridges; velvety; white when young, later yellowish.

Flesh White or creamy, thin (1-2mm; 1/32-1/16in) waxy brittle, almost odourless and tasteless.

Spore deposit Pale ochre.

Spores 20-24×12-14μm, elliptical; smooth; colourless.

Habitat It grows from the end of March in meadows, woods and damp, grassy places; it may appear also in gardens, on compost and in ruins. It differs from *Morchella esculenta* in its shape and the colour of its cap.

Use This tasty edible mushroom can be used for soups, sauces and all sorts of meat dishes.

Remarks A very similar mushroom is the Morel *Morchella elata* (Fr.): its cap is cylindrical, conical, 30-600mm (1⅛-2¼in) high, with sharp longitudinal ribs. Crosswise partitions create irregularly-placed dimples between the ribs. The dimples vary in colour from olive-brown through brownish-green to a very dark brown; the ribs and crosswise partitions are usually pale olive-yellow. The stipe is cylindrical, 50-150mm (2-6in) long, first off-white, later yellow to yellowish-brown. It grows in April and May in deciduous woods, gardens and greenhouses. It is edible.

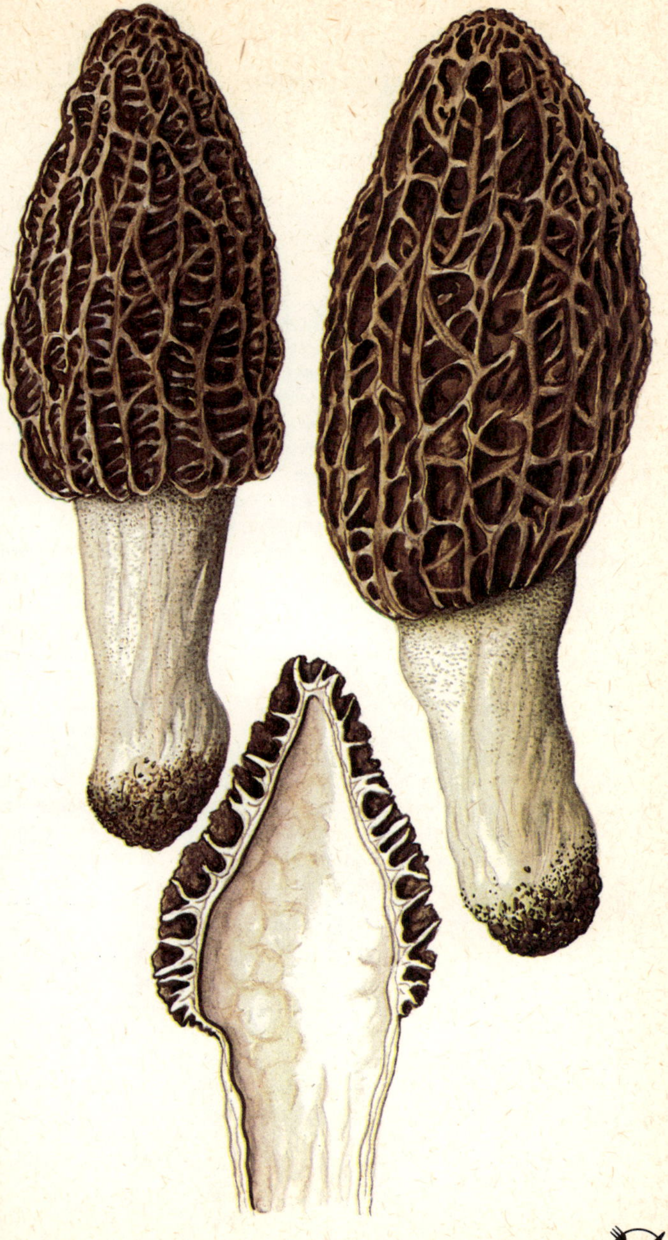

FALSE MOREL **Poisonous**

Gyromitra esculenta (Pers. ex Pers.) Fr.

Cap	Red-brown, chestnut-brown or dark brown with purple shading, 25-80mm (1-3⅛in) in diameter; irregularly shaped; its surface is curved and folded like a brain, partially attached to the stipe, partially having lobed protrusions; inside it is hollow and white.
Stipe	Irregular, cylindrical, hollow, brittle, often double or treble, 30-70mm (1⅛-2¾in) long, 15-35mm (⅝-1⅜in) thick; its surface is dimpled and wrinkled, velvety, minutely hairy, whitish, yellowish or flesh-coloured.
Flesh	White, thin (1-2mm; 1/32-1/16in) brittle and crumbling, very mild taste, little smell.
Spore deposit	White.
Spores	$18\text{-}20 \times 9\text{-}12\mu m$, elliptical; smooth surface with two small oil drops, colourless.
Habitat	It grows from the beginning of March until the end of April in coniferous woods. It is most frequent in pine woods with sandy soil.
Use	This fungus is highly poisonous.
Remarks	The poisonous *Gyromitra esculenta* may be mistaken for the edible Giant Giromitra *Neogyromitra gigas* Krombh. Imai); its fruiting body is usually bigger; it has an orange-yellow cap and it grows in deciduous woods.

†

CAULIFLOWER FUNGUS **Edible**

Sparassis crispa (Wulf.) ex Fr.
Syn.: Masseola crispa (Wulf. ex Fr.) O. Kuntze
Sparassis ramosa Schaeff. ex Schroet.

Fruiting body 50-200mm (2-8in) tall, 60-300mm (2¼-12in across), but it can be even bigger, and sometimes reaches a weight of 6kg (13lb). Its shape is irregularly spherical or elliptical, with a thick fleshy base from which leafy branches grow and develop; the branches are partially fused; both sides are covered by hymenium. The edges of the wavy or curled lobed leaves are toothed. At first they are whitish, later yellowish and when old they become orange-yellow to yellowish-brown. The base of the stipe is almost black.

Flesh Waxy, flexible, white, with a pleasant smell, tastes somewhat like nuts.

Spore deposit Yellow to orange-yellow.

Spores 6-7 × 4-5μm, short, elliptical; smooth; light-yellow.

Habitat It grows from August to October in coniferous woods, usually close to the roots of pine trees.

Use A delicious edible mushroom, suitable for soups, for frying and for preserving.

Remarks *Sparassis crispa* (Wulf.) ex Fr. looks very much like *Sparassis laminosa* but the latter's straw-yellow leafy branches are thicker, flat and smooth. It grows at the end of summer and in autumn in deciduous woods, usually near the roots of oak trees. It is tasty, like *Sparassis crispa*, but much rarer.

CHANTARELLE **Edible**

Cantharellus cibarius Fr.

Cap When young it is yolk-yellow, later it becomes paler; it is usually bare and dull; 10-70mm (6-2¾in) in diameter; at first it is slightly convex, then almost flat. When grown large it may be funnel-shaped, with a waxy, lobe-like incurved margin. The hymenophore on the underside of the cap is composed of branching veins linked by partitions. The veins may stand about 3mm (⅛in) above the surface; they are rather thick, bright yellow in colour, all running towards the stipe.

Stipe Irregular, cylindrical, narrowed at its base, extended at its summit, solid, 30-60mm (1⅛-2¼in) long, 8-25mm (⅓-1in) thick, with a yellow surface, smooth and dry.

Flesh Thick, pale yellow, with a pleasant spicy smell and taste.

Spore deposit Pale yellow.

Spores 8-11×5-6μm, elliptical; smooth; colourless, with one or more oil drops.

Habitat It grows from the end of May till the end of November in mixed woodland among moss and fallen leaves.

Use It can be cooked in many ways. When young and pickled it is a real delicacy which is served with meat. It is rather tough, therefore not easily digested. But it is an important mushroom, because it grows in great quantities and is not usually damaged by insects.

Remarks The inedible *Hygrophoropsis aurantiaca* (Wulf. ex Fr.) R. Maire is very similar at first sight to the chantarelle. Its cap is thin and fleshy; at first it is orange, later becomes pale; its gills are dense, thin and orange. Its flesh is soft, pale orange and tastes bitter. It grows from September to November in mixed woodland.

HORN OF PLENTY **Edible**

Craterellus cornucopioides (L. ex Fr.) Pers.
Syn.: Cantharellus cornucopioides L. ex Fr.

Fruiting body The cap is 30-80mm (1⅛-3⅛in) in diameter; it is trumpet- or funnel-shaped with a folded margin which when fully grown is usually wavy or lobe-like; its inner part is at first dark brown; later it goes very dark grey, even black; scaly; on the outside it is usually greyish-brown; at first it is smooth, later wrinkled. Outside it is completely covered with hymenium, so a mature fungus is white and covered with bloom. The stipe body is 50-120mm (2-4¾in) long, hollow, narrowing towards the base; at the top it gradually merges into the cap.

Flesh Very easily broken, slightly leathery; at first it is greyish-black, later deep black. Its taste is mild, it smells pleasant. When dried it is black and easily broken.

Spore deposit White.

Spores 12-15×7-8.5μm, egg-shaped or elliptical; smooth; colourless.

Habitat It grows from August to November in deciduous, leafy woods in great clusters. It is most common on the lower slopes of mountains.

Use A very tasty mushroom in spite of its unattractive appearance. It is good for soups and sauces. Dried and crumbled it can be used as seasoning. Its smell is reminiscent of truffles and it is a good flavouring for some stews.

Remarks Its trumpet-like funnel cannot be mistaken for any other mushroom. The most like it is *Craterellus sinuosus* (Fr.) Fr. which differs from it by having a more incurved and curly cap and by its stipe, which is hollow only as far as, but not into, the cap. Apart from that it is paler, usually in shades of brown and orange-yellow. It grows from August to October in deciduous woods in large bunches. It is not edible.

SCALY HYDNUM **Edible**

Hydnum imbricatum L. ex Fr.
Syn.: *Sarcodon imbricatum* (L. ex Fr.) P. Karst.

Cap Brown to grey-brown; when young it is flat and slightly convex, later it develops a depression in the centre, even becoming funnel-shaped. The margin is folded at first, then it unfolds to become incurved. The cap is 50-200mm (2-8in) in diameter and its surface is covered with thick, protruding scales. The spines on the underside of the cap are dense and easily broken; when young they are whitish, later they become grey or grey-brown; when mature they are brown; all spines point to the stipe.

Stipe Irregular, cylindrical, often thicker at its base, solid, 30-60mm ($1\frac{1}{8}$-$2\frac{1}{4}$in) long, 10-30mm ($\frac{3}{8}$-$1\frac{1}{8}$in) thick with a smooth surface, grey-brown or brown.

Flesh At first it is whitish, later greyish-brown; flexible and juicy when young, when ripe it develops a hard dry skin. It tastes spicy, and smells piquantly pleasant.

Spore deposit Brownish-grey.

Spores $5\text{-}7 \times 5\text{-}6\mu m$, oval with an angular rugged surface; pale yellow.

Habitat It grows from August to November in coniferous woods, almost always in large clumps.

Use Young mushrooms are suitable for picking while they are still juicy and pliable. Old ones are skinny, tough and bitter. Dried and crumbled, they are good for seasoning soups and sauces.

Remarks Bitter Hydnum (*Hydnum scabrosum*) Fr. can easily be mistaken for *Hydnum imbricatum*. *Hydnum scabrosum* differs from *Hydnum imbricatum* in the colour of the cuticle of its cap. When young this is pinkish-purple and smooth; the mature mushroom is usually orange-yellow to brown and covered with cracks. It has no scales. The base of its stipe is greenish-black. Its flesh is bitter, and it is not edible.

WOOD HEDGEHOG **Edible**

Dentinum repandum (L. ex Fr.) S. F. Gray
Syn.: *Hydnum repandum* L. ex Fr.

Cap When damp it is a creamy-yellow, pale orange or yellowish-brown; when dry it is almost white. At first convex, it becomes flat, usually with the margin curved downwards; 30-150mm (1⅛-6in) in diameter. The spines on the underside of the cap are close together and extend down the stipe; when young they are short and flexible, pale yellow, later they are almost 8mm (⅓in) long, of the same colour as the cap and brittle.

Stipe Cylindrical, 20-70mm (¾-2¾in) long, 5-20mm (3/16-¾in) thick; its surface is smooth, dry and whitish or creamy, in rare cases yellowish-brown.

Flesh Tough and rather brittle, white or yellowish-white; it does not change its colour when cut or sliced. It tastes slightly sour and has a pleasant smell.

Spore deposit White.

Spores 7-10 × 7-8μm, nearly spherical or egg-shaped; smooth or finely reticulated surface; colourless.

Habitat Grows from July to November in mixed woodland, in large clumps sometimes forming circles. It is most common in the foothills of mountains.

Use This mushroom is edible, but due to its relatively hard flesh it is not easy to digest. It is most suitable for soups and pickling mixed with other kinds of mushrooms.

Remarks At first sight it is very similar to the Chantarelle but it differs from it in the spines that grow on the underside of its cap instead of the gills which the chantarelle has. Another fungus similar to *Dentinum repandum* is *Dentinum repandum f. rufescens* Fr., which is usually smaller and has a rusty-orange or rich orange cap. It grows in the same place and at the same time and is edible.

OYSTER MUSHROOM **Edible**

Pleurotus ostreatus (Jacq. ex Fr.) Kumm.

Cap Greyish-blue, greyish-brown or grey, 50-200mm (2-8in) across. When young it is convex, growing flatter in outline, hemispherical or shell-shaped, with a smooth, bare surface. Its gills are at first white, then greyish, 5-15mm (3/16-⅝in) wide and close together, extending far down the stipe.

Stipe Off-centre or to one side; irregularly cylindrical, solid, 20-50mm (¾-2in) long, 8-20mm (⅓-¾in) thick, with a white surface, whitish and minutely hairy at its base.

Flesh White; when young it is soft and juicy, later it becomes tough; its taste is mild and its smell pleasant.

Spore deposit White with a slight purple tinge.

Spores 8-13 × 3-4μm, cylindrical; smooth; colourless.

Habitat It grows from October to December, usually in great clumps on deciduous tree trunks and stumps. If the weather is not very frosty it can grow all winter up to March.

Use *A good edible mushroom, suitable for soups and all kinds of pickling. It is best to eat the young mushrooms while they are soft.*

Remarks There is no poisonous mushroom that could be mistaken for the Oyster Mushroom. There are only few sorts of *Pleurotus* that look similar to it. All of them are edible and there is no danger of poisoning. From May to October they grow on trunks of deciduous trees. The most common is the Branched Oyster Fungus (*Pleurotus cornucopiae* Paul. ex Fr.) which grows in large clusters on elm trees. Its cap is funnel-shaped, whitish, pale yellow, or yellowish-brown. The gills are white, extending down the stipe; at the base they are reticulated and thus linked to each other. It is a tasty edible mushroom; it can be prepared in the same way as the Oyster Mushroom.

MARCH MUSHROOM **Edible**

Hygrophorus marzuolus (Fr.) Bres.

Cap 40-100mm (1½-4in) wide, first convex, later flat, often irregularly incurved; fleshy, symmetrical or lop-sided. When young its cuticle is off-white, later mother-of-pearl grey to black, matt and dry; at first slightly bloomed, later smooth. The gills are 2-6mm (1/16-¼in) wide, widely spaced, rather thick, attached a short way down the stipe, white or off-white when young; with age they become greyish or grey, the margin being pale grey. Irregularly cylindrical, solid, thinner at its base, 30-60mm (1⅛-2¼in) long, 15-40mm (⅝-1½in) thick; it is often curved to some extent; its surface is off-white or greyish, silky and fibrous, scaly at its apex under the cap.

Flesh Close-textured; when young it is vivid white, later it develops a slight greyish shade especially under the cap cuticle and in the stipe; it is odourless and tastes mild.

Spore deposit White.

Spores 7-9×4-5μm, egg-shaped or elliptical; smooth; colourless.

Habitat It grows from March to the beginning of May in coniferous woods; often it can be found under the remains of snow, exceptionally in deciduous woods. It is most frequent on lower mountain slopes. It is quite difficult to find because of its dull, colourless cap.

Use An edible mushroom, popular and sought after because it grows early in the spring when few sorts of mushrooms are growing. It can be used for soups and meat dishes.

Remarks No poisonous fungus can be mistaken for *Hygrophorus marzuolus.* Some confusion may occur with some inedible sorts of *Russula* which have a greyish-brown cap, but this is unlikely because Russula starts to grow at the beginning of the summer while *Hygrophorus marzuolus* is over by the beginning of May.

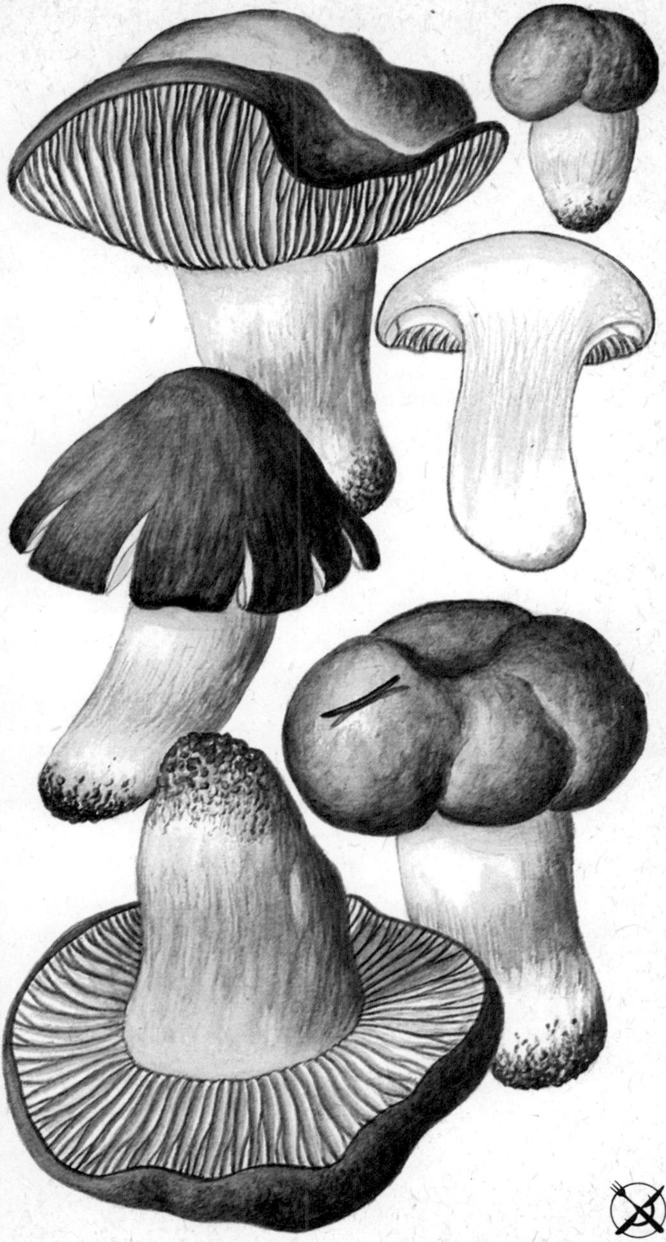

LYOPHYLLUM DECASTES **Edible**

Lyophyllum decastes (Fr. ex Fr.) Sing.

Cap Yellow-brown to rust-brown, with embedded radiating fibrils and a silky sheen; when young it is hemispherical and convex; later it becomes flat and undulating; 50-100mm (2-4in) in diameter. It is usually symmetrical, in rare cases lop-sided. Fruiting bodies growing in shade usually have lighter-coloured caps.

The gills are at first white, later greyish-cream, 6-8mm (¼-⅓in) wide; moderately close together; notched where they join the stipe.

Stipe Solid, rarely off-centre, 70-130mm (2¾-5⅛in) long, 10-20mm (⅜-¾in) thick, with white, longitudinal minute fibrils, and having a silky sheen.

Flesh Close-textured, pliable, white, does not change its colour when cut. It has a mild flavour and is odourless.

Spore deposit White.

Spores 5.5-6.5μm, spherical, smooth; colourless.

Habitat It grows in large clumps from September to the beginning of December in meadows and deciduous woods; often found in gardens and parks.

Use A good edible mushroom, very solid; it is almost always prolific (often 50-100 fruiting bodies in one clump). It is suitable for soups and pickling. Only the caps are eaten.

Remarks No poisonous fungus can be mistaken for *Lyophyllum decastes*. Some related kinds could be taken for it, but all are edible, for example, *Lyophyllum fumosum (Pers. ex Fr.) P. D. Orton* and *Lyophyllum connatum* (Schum. ex Fr.). Sing. *L. fumosum* grows in clumps at the same time and in similar places to *L. decastes*, from which it differs in the darker colouring of its cap, and the larger size of its fruiting body. *L. connatum* grows in clusters from August to October in mixed woodland. *L. connatum* is all white.

ST GEORGE'S MUSHROOM **Edible**

Calocybe gambosa (Fr.) Donk
Syn.: *Tricholoma georgii* (L. ex Hook.) Quél.

Cap At first creamy, when old it goes orange-yellow to yellowish-brown; smooth, bare and dry. The cap of the young mushroom is a rounded cone or bell shape; later it becomes convex and when old it spreads out flat; 60-120mm (2¼-4¾in) in diameter. The gills are at first white to cream; then they become pale orange-yellow; 5-10mm (3/16- ⅜in) wide, very closely set and slender, almost always notched where they join the stipe. The margin of the gills later becomes toothed and torn.

Stipe When young it is pot-bellied or club-shaped, then cylindrical; 40-90mm (1½-3½in) long, 15-35mm (⅝-1⅜in) thick, whitish or pale cream; its base is usually orange-yellow or rust. Its surface is streaked with fibres; immediately under the cap the stipe may be slightly scaly.

Flesh Close-textured, juicy, white, does not change its colour when cut, tastes pleasant and smells like freshly ground flour.

Spore deposit White.

Spores $5\text{-}7 \times 3\text{-}4.5\mu m$, egg-shaped or compact ellipse; smooth; colourless.

Habitat It grows from the end of April to the beginning of June, especially in meadows, pastures, on the edges of deciduous woods and — rarely — in gardens. In mountain regions with harsher climatic conditions it arrives later, from the beginning of June to the first half of July. It can be found alone or in small clumps, also in circles.

Use This tasty edible mushroom is good for soups and pickling. When dried it can be used for flavouring meat dishes.

Remarks The highly poisonous *Inocybe patouillardii* Bres., which is white when young, may easily be mistaken for the St George's Mushroom, but the young *Inocybe patouillardii* turns red when it is bruised or scratched; older specimens are easily distinguished from the St George's Mushroom by the fibrous red-brown cap and their brown gills.

HONEY FUNGUS **Edible**

Armillariella mellea (Vahl. ex Fr.) P. Karst.
Syn.: *Clitocybe mellea* (Vahl.)

Cap Usually honey-yellow or rusty-brown, rarely yellowish-green or dark brown, covered by brown scales which are closest-set and darkest in the centre of the cap. When mature the scales partially disappear. When young the cap is hemispherical, later convex, becoming flatter at maturity; it often has a rounded hump in its centre; the cap is 40-100mm (1½-4in) in diameter. The gills are first off-white, later yellowish with brown spots; when mature they are usually white, powdered with spores. They are 4-8mm (5/32-5/16in) wide, moderately close together, thin, extending slightly down the stem.

Stipe Cylindrical, usually swollen at the base, with a soft, spongy interior, 50-200mm (2-8in) long, 10-25mm (⅜-1in) thick, yellow or orange-brown, with fine fibrous grooves or covered with scales. When young the margin of the cap is connected with the stipe by a membranous, soft scaly veil. When older it often leaves a loosely-hanging ring on the stipe.

Flesh White; retains its colour when sliced; it is usually soft in the cap but hard and fibrous in the stipe. It tastes bitter although it is odourless.

Spore deposit White.

Spores $7\text{-}9.5 \times 5\text{-}6.5\mu m$, cylindrical or elliptical; smooth; colourless.

Habitat It grows from the second half of September to the end of October on the trunks of coniferous and deciduous trees, in large clusters.

Use It is an edible mushroom of average quality but not easily digestible. If not well cooked it may cause stomach troubles: it may even have a slight poisonous effect on some people. It is most suitable for pickling. For cooking, use only the caps. Foresters regard this fungus as an enemy because it is parasitic on the trunks of various trees, mainly spruces.

Remarks *Armillariella tabescens* is very similar to Honey Fungus. It differs from Honey Fungus by lacking a ring on its stipe; it grows only on the trunks of deciduous trees. It is edible, too.

TRUMPET AGARIC **Edible**

Clitocybe geotropa (Bull. ex St-Am) Quél.

Cap Yellowish or brownish-yellow, smooth. The cap of the young mushroom is at first white and convex with a downward curving margin; it has quite a large central hump which stays the same size as the mushroom develops and the cap gradually deepens into a funnel shape; it is 80-250mm (3⅛-10in) in diameter. The gills are first white, later creamy-yellow, 3-10mm (⅛-⅜in) wide, closely set, slender and pointing down the stipe.

Stipe Cylindrical, swollen at its base, solid, 100-300mm (4-12in) long, 10-40mm (⅜-1½in) thick, pale orange-yellow, smooth and slightly fibrous; white and velvety at its base.

Flesh Close-textured, white; keeps its colour when cut; it has a mild taste and is odourless.

Spore deposit White.

Spores $6\text{-}7 \times 5\text{-}6\mu m$, almost spherical; smooth; colourless.

Habitat It grows from August to November in meadows and deciduous woods. Appears usually in small groups or in circles.

Use An edible mushroom, suitable mainly for pickling in combination with other more tasty mushrooms. Only the caps of young mushrooms are used because the stipe is fibrous, tough and hard to digest.

Remarks No poisonous fungus can be mistaken for *Clitocybe geotropa*. A similar-looking fungus is *Leucopaxillus giganteus* (Sibth. ex Fr.) Sing, but this has no hump in the centre of its cap. It is edible, but its flavour is less pleasant than that of *Clitocybe geotropa*.

CLOUDED AGARIC **Edible**

Clitocybe nebularis (Batsch ex Fr.) Kumm.

Cap When young it is hemispherical, and convex; then it becomes flat and extended, sometimes depressed in the centre; its margin curves inwards at first; it unfolds later, often becoming wavy; it is 50-150mm (2-6in) in diameter. The cuticle looks darker when moist; in damp weather deep grey to grey-brown; in dry weather pale grey or a light grey-brown. When young its whole surface has a whitish bloom, which in older mushrooms remains in the centre of the cap; it can be easily wiped off.

The young gills are off-white; when older they turn creamy-yellow, 3-7mm (1/8-9/32in) wide, closely set and slender; pointing slightly down the stipe.

Stipe Cylindrical or club-shaped, rounded at the base, solid, 60-100mm (2¼-4in) long, 15-30mm (5/8-1 1/8in) thick. Its surface is smooth or finely ridged with fibres; off-white, greyish or slightly brownish at some places.

Flesh Tough, thick, white; keeps its colour when cut. Its smell and flavour are very variable. Sometimes it smells like freshly ground flour, at others it has an unpleasant odour, while the flavour may be spicy, sweet or rather sour.

Spore deposit White.

Spores 6-8 × 3-4μm, elliptical; smooth; colourless.

Habitat It grows from the end of September to December in mixed woodland in great quantities, often in circles.

Use Even though it is an edible fungus, it is not very tasty. It is hard to digest, highly scented and for some people it causes stomach trouble. Only young mushrooms are used. It is best pickled in combination with other mushrooms.

Remarks No poisonous fungus can be mistaken for Clouded Agaric.

CLITOCYBE DEALBATA **Poisonous**

Clitocybe dealbata (Sow. ex Fr) Kumm.

Cap At first it is convex with an inward curving margin; later often depressed in the centre; 20-45mm (¾-1¾in) in diameter. It is either milk white or very pale grey, smooth; when old the surface is sometimes cracked in concentric circles. In damp weather it is slightly sticky; when dry it has a silky sheen. The gills when young are milk white, later becoming pale grey; when mature it has creamy, closely set gills — 2-3mm (1/16-3/16in) wide running down the stipe.

Stipe Cylindrical, solid, 20-40mm (¾-1½in) long, 4-6mm (5/32-¼in) thick, white, greyish at its base and streaked with longitudinal fibres.

Flesh Thin, flexible, white, retains its colour when sliced. It has a mild flavour, and smells like freshly-ground flour or newly-sawn logs.

Spore deposit White.

Spores $4\text{-}6 \times 2.5\text{-}3.5\mu m$, elliptical; smooth; colourless.

Habitat It grows in meadows and pastures or on the edges of deciduous woods from August to November, singly or in small clumps.

Use It is highly poisonous.

Remarks *Clitocybe cerussata* (Fr.) Kumm., differs from *Clitocybe dealbata* by its larger fruiting body. Its cap is 50-120mm (2-4¾in) wide; when young it is convex, later flat or almost funnel-shaped, white and with a sheen. Its gills are dense and grow a little way down the stipe. At first they are white, later they become off-white to creamy. The stipe is 50-100mm (2-4in) long, 7-12mm (9/32-½in) thick and whitish-brown at the base. Its flesh is white, it has a mild taste and no smell; it grows from August to October in woodland. It is also very poisonous.

†

WOOD BLEWITS **Edible**

Lapista nuda (Bull. ex Fr.) W. G. Smith
Syn.: *Tricholoma nudum* (Bull. ex Fr.) Kumm.

Cap	When young it is deep purple, later pale purple; at maturity it turns pale and is usually yellowish-brown, in dry weather orange-yellow. At first it is convex, becoming flat, smooth and bare. It reaches 60-150mm (2¼-6in) in diameter. The gills when young are deep purple, then orange-brown to flesh-brown; 6-10mm (¼-⅜in) wide, close together, slender and notched at the stipe.
Stipe	Cylindrical, rarely also club-shaped, solid; 50-100mm (2-4in) long, 12-30mm (½-1⅛in) thick; the surface is at first deep purple, later almost greyish-white with longitudinal fibrils; at the base it is a distinctive purple, velvety, and covered with leaf-like structures.
Flesh	At first tough and purple; a mature fungus gradually softens and turns to a pale creamy shade. It has a mild, pleasant flavour and is odourless.
Spore deposit	When fresh it is white tinged with pink; when dried the pink is deeper.
Spores	7-9×4-5.5μm, elliptical; warty; colourless.
Habitat	It grows from the end of September to December in mixed woodland, singly or in groups. It is frequent in fields, lawns and gardens.
Use	As one of the most tasty autumn mushrooms, it is most suitable for pickling and in meat dishes. It is not advisable to eat it raw because it contains a substance which dissolves the haemoglobin in red blood cells. This haemolysin is neutralised by cooking.
Remarks	At first sight Wood Blewits looks like the inedible Goaty Smell Cortinarius (*Cortinarius traganus*) (Fr. ex Fr.) Fr., the flesh of which is yellow and bitter with a repulsive smell.

BLEWITS

Edible

Lepista saeva (Fr.) P. D. Orton
Syn.: *Tricholoma personatum* (Fr. ex. Fr.) Kumm.

Cap Pale cream, yellow or pale orange-yellow, sometimes with a brownish tinge. When young it is usually hemispherical; later it becomes convex; the mature fungus is flat, smooth and bare; 60-150mm (2¼-6in) in diameter.

The gills are at first off-white, then pale cream; mature gills are pale orange-yellow, 7-10mm (9/32-⅜in) wide, closely set, slender and notched where they join the stipe.

Stipe Cylindrical, swollen at the base, solid; 30-70mm (1⅛-2¾in) long, 15-30mm (⅝-1⅛in) thick; it is beige, covered with dark purple longitudinal fibrils, which make it all look purple.

Flesh Tough, thick, white or pale cream, right under its surface sometimes slightly purple; when cut it retains its colour. The flavour is mild and pleasant, and it is odourless.

Spore deposit White, with a pinkish tinge.

Spores 5.5-7 × 4-5 μm, elliptical; warty; colourless.

Habitat It grows from October to December in meadows and pastures, rarely in open deciduous woods.

Use As a good edible mushroom it is suitable for pickling, in meat dishes, stewed with onions and in omelettes.

Remarks *Lepista irina* (Fr.) Bigelow is very similar to Blewits but has thinner gills, and a pale orange-yellow stipe without any purple shading. Its pleasant smell is reminiscent of violets. It grows fairly abundantly from October to December in meadows. It is edible.

MAN ON HORSEBACK, FIRWOOD AGARIC

Edible

Tricholoma flavovirens (Pers. ex Fr.) Lund.
Syn.: *Tricholoma equestre* (L. ex Hook) Kumm.

Cap Brownish or greenish-yellow, darkest at the centre, smooth; in damp weather it is sticky, almost always with soil clinging to it. When young it is convex; later it becomes flat; often with a wavy margin, it is 40-120mm (1½-4¾in) in diameter.

The gills are deep yellow, notched at the stipe, 5-12mm (3/16-½in) wide, slender and moderately close together.

Stipe Cylindrical, solid, 30-90mm (1⅛-3½in) long, 10-30mm (⅜-1⅛in) thick, with longitudinal fibrils at its surface; later bare; yellow or greenish-yellow.

Flesh Firm and brittle; white, but under the cuticle of the cap it is a bright lemon yellow; it retains its colour when cut. It smells like freshly ground flour and tastes sweet and nutty.

Spore deposit White.

Spores 6-8 × 3.5-5μm, egg-shaped or elliptical; smooth; colourless.

Habitat It grows from October to December in pine woods, singly or in large groups. As the fruiting body develops almost entirely underground it is quite difficult to find because often only part of its cap can be seen among moss and pine-needles. It is very common in sandy soil.

Use This delicious edible mushroom is very good for soups and in meat and egg dishes.

Remarks There is a striking resemblance between *Tricholoma flavovirens* and the slightly poisonous *Tricholoma sulphureum* (Bull. ex Fr.) Kumm. *T. sulphureum* is smaller, its gills are more spaced apart and its flesh is yellow all over. It has a repulsive tarry smell. It grows from July to October in woodland.

DINGY AGARIC **Edible**

Tricholoma portentosum (Fr.) Quél.

Cap Pale or dark grey, often with a slight purple tinge; 40-100mm (1½-4in) in diameter. When young it is conical; later it becomes flat and spread out, with a wavy or lobed margin, often with a slight hump at its centre. The cuticle of the cap is smooth and streaked with fibrils. When mature it usually splits. In damp weather it is slimy.

The gills are at first white, later greyish with a lemon-yellow fringe; 5-10mm (3/16-⅜in) wide, moderately close together, slender and notched at the stipe.

Stipe Cylindrical, solid, 50-100mm (2-4in) long, 10-20mm (⅜-¾in) thick, with a smooth, bare or minutely scaled surface, white with a yellowish or greyish tinge.

Flesh White or grey, often yellowish; rather thin; it tastes pleasant and smells like freshly-ground flour.

Spore deposit White.

Spores 6-7 × 4-5μm, egg-shaped or elliptical; smooth; colourless.

Habitat It grows from October to December in pine woods together with *Tricholoma flavovirens*, almost always in large groups.

Use This very tasty mushroom is suitable for soups, frying, pickling, and drying.

Remarks A very similar mushroom is Grey Agaric *Tricholoma terreum* (Schaeff. ex Fr.) Kumm. Its cap has a fibrous and scaly surface, and finer, more widely-spaced gills. It grows from the end of September until winter in mixed woodland. It is edible but not so tasty as *Tricholoma portentosum*. The poisonous *Tricholoma virgatum* (Fr.) Kumm can also be easily mistaken for *T. portentosum*, but it is slimmer, has a grey cap with a conical hump at its centre, greyish gills and a peppery taste.

TIGER TRICHOLOMA **Poisonous**

Tricholoma pardalotum Herink et Kotl.
Syn.: *Tricholoma pardinum* Quél

Cap	Greyish-brown or dark grey; covered with minutely fibrous scales which are most close set at the centre. When young it is bell-shaped; later it spreads out flatter; 60-120mm (2¼-4¾in) in diameter. The gills are at first white, often with a greenish tinge; the mature gills are usually pale yellow but never notched where they joined the stipe; moderately close together, 8-12mm (⅓-½in) wide.
Stipe	Cylindrical or club-shaped, solid, 40-80mm (1½-3⅛in) long, 20-35mm (¾-1⅜in) thick; the young fungus has a finely fibrilar stipe, but later it becomes almost smooth and white; its bottom half is ochre, its base a rusty yellow.
Flesh	White; it retains its colour when cut; it has a mild flavour and smells faintly pleasant.
Spore deposit	White.
Spores	8-10 × 6-7μm, egg-shaped or elliptical; smooth; colourless.
Habitat	It grows from August to October in mixed woodland, singly and scattered. It is quite rare. Its fleshy well-developed fruiting bodies are attractive but dangerous.
Use	This is a very poisonous fungus, causing extremely painful illness.
Remarks	At first sight the edible *Tricholoma orirubens* Quél., appears similar: it has finer, olive-grey scales on its cap and, when mature, has pink gills. It grows from September to November in deciduous woods, usually under beech trees.

†

WINTER MUSHROOM, VELVET SHANK **Edible**

Flammulina velutipes (Curt. ex Fr.) Sing.
Syn.: *Collybia velutipes* (Curt. ex Fr.) Kumm.

Cap Yellow-orange, orange or yellowish-brown, smooth, bare; at its margin the gills are partly visible as fine lines through the cap. In damp weather it is very slimy. When young it is hemispherical or convex, flattening as it matures; 15-70mm (5/8-2¾in) in diameter.

The gills are at first pale cream, then orange-yellow, widely-spaced, rounded near the stipe and 5-12mm (3/16-½in) wide.

Stipe Usually centred but rarely slightly off-centre, cylindrical and hollow, 50-100mm (2-4in) long, 3-12mm (1/8-½in) thick; reddish-yellow or honey-yellow and bare at its base; the rest of it is usually very dark brown and velvety.

Flesh Thin, white or creamy; retains its colour when cut; it has a mild flavour and a faint, pleasant smell.

Spore deposit White.

Spores 6-9×4-4.5μm, cylindrical or elliptical; smooth; colourless.

Habitat It grows from October to December on the boles and trunks of deciduous trees, almost always in thick clusters. It is most common on poplars and willows. If the winter is mild it grows until March.

Use This edible mushroom is good for soups and pickling. In spite of its slimy cap it is popular because it grows in winter when other mushrooms have died back.

Remarks No other mushroom can be mistaken for *Flammulina velutipes*, except perhaps some sorts of *Hypholoma* or the Changeable Agaric, or *Kuehneromyces mutabilis* (Fr.) Sing and A. H. Smith. All these fungi, growing in large clumps, have dark coloured gills.

FAIRY RING MUSHROOM, CLOVER WINDLING

Edible

Marasmius oreades (Bolt. ex Fr.) Fr.

Cap	Smooth; in damp weather it is reddish-yellow or brownish-yellow; in dry weather it is orange-yellow to pale cream; at its margin the gills show through as vertical lines. When young it is a rounded cone or a hemisphere, later convex; it flattens towards maturity, often having a slight hump in the middle; 20-50mm (¾-2in) in diamter.
Stipe	Cylindrical, solid, 40-70mm (1½-2¾in) long, 3-4mm (about ⅛in) thick, flexible, not easily broken, pale orange-yellow; when young it is minutely hairy, later it becomes bare.
Flesh	Thin, whitish; keeps its colour when cut; it smells pleasant and tastes delicious.
Spore deposit	White.
Spores	8-10×4-5μm, egg-shaped or elliptical; smooth; colourless.
Habitat	It grows from the first half of May to the end of November, often in large circles, in meadows, gardens, in the grass bordering field paths, and at the edge of woodland. It grows in profusion after heavy rains in lowlands and at high altitudes.
Use	A very good edible mushroom; adds an excellent flavour to soups.
Remarks	Wood Agaric or Oak-living Collybia — *Collybia dryophila* (Bull. ex Fr.) Kumm., is very similar to the Fairy Ring Mushroom at first sight. It grows freely from May to December in deciduous and coniferous woods. It is edible, but its flavour is not so fine as that of the delicious Fairy Ring Mushroom. Its cap is brownish-yellow to light brown in damp weather; in dry weather it is pale cream, smooth and bare. When young it is bell-shaped, later flat; it is 20-60mm (¾-2¼in) in diameter. Its gills are very tightly packed; when young they are white, when mature they become creamy-yellow. The stipe is cylindrical and hollow. It is 30-70mm (1⅛-2¾in) long, 3-5mm (⅛-3/16in) thick, easily broken; yellowish-brown or reddish-brown. The flesh is very pale and thin; it has a mild taste but smells unpleasant. The powder is white. Spores are 6-7×3-4μm, elliptical, smooth and colourless.

SHEATHED AGARIC, TAWNY GRISETTE **Edible**

Amanita fulva (Schaeff) ex Pers.

Cap It is 30-100mm (1⅛-4in) in diamter; at first it is bell-shaped, later it becomes convex, almost flat, with a hump at its centre; it has fine vertical lines along the margin; yellowish-brown to reddish-brown, bare; in damp weather it is sticky and slimy.

The gills are 5-10mm (3/16-⅜in) wide; moderately close together, white or creamy, not attached to the stipe; rounded near the margin of the cap.

Stipe 80-150mm (3⅛-6in) long, 8-12mm (⅓-½in) thick, cylindrical, thinning towards the upper end and thickening towards the base; when mature it is hollow, with a white, smooth, bare surface and no ring; at its base it has a white or orange-yellow free volva reaching almost a quarter of the way up the stipe.

Flesh White, thin and brittle; it has a sweet flavour and no smell.

Spore deposit White.

Spores 9-12μm, spherical; smooth; colourless.

Habitat It grows in birch woodlands from spring to autumn.

Use It is an edible mushroom, but not very substantial. It breaks easily when picked. If eaten in large quantities it may cause slight poisoning.

Remarks There is an affinity between *Amanita fulva* (Schaeff) ex Pers. and the Common Grisette (*Amanita vaginata*) (Bull. ex Fr.) Vitt. but *A. vaginata* has a grey cap. It grows from July to October in both coniferous and deciduous woods, mainly beech, and is also edible.

FLY AGARIC

Poisonous

Amanita muscaria (L., ex Fr.) Hook.

Cap Scarlet, almost carmine red, covered with white fragments of its universal veil. When young it is almost spherical, later it becomes convex and flat, with thin lines around the margin; 100-200mm (4-8in) in diameter. Its gills are white or creamy yellow, 6-12mm (¼-½in) wide, close together, not attached to the stipe and rounded near the cap margin.

Stipe Cylindrical, with a bulbous base; when young it is solid; the mature stipe becomes hollow; 80-200mm (3⅛-8in) long, 10-35mm (⅜-1⅜in) thick; it is easily separated from the cap. The upper section of the stipe is covered with a thick ring. Its surface is white, scaly or flaky, and the basal bulb bears a few concentric rings of warts.

Flesh White; under the cap cuticle it is bright yellow; its retains its colour when cut. It tastes slightly sweet, and has little smell.

Spore deposit White.

Spores $10\text{-}12 \times 6\text{-}7\mu m$, elliptical; smooth; colourless.

Habitat It grows from August to November in coniferous and deciduous woods, singly or in large groups. It is most common in spruce woods in mountainous and hilly regions. It is one of the most beautiful fungi and is the traditional 'toadstool' of fairy-tale illustrations.

Use This poisonous fungus is not very dangerous because it is so well known. Its harmful effects are often exaggerated, but it is much less dangerous than other sorts of poisonous agarics. Some of the toxic components of Fly Agaric produce hallucinations; certain Asian tribes used to prepare an intoxicating drink from it which they consumed on festive occasions. A poison for flies used to be prepared from this fungus — hence its popular name.

Remarks The Royal Amanita *Amanita regalis* (Fr.) Michael is related to the Fly Agaric but differs from it by its brown cap and yellowish stipe; it grows in summer and autumn in spruce woods on mountains; it is also poisonous.

†

THE BLUSHER

Edible

Amanita rubescens (Pers. ex. Fr.) S. F. Gray

Cap When young it is whitish-pink; later it becomes flesh-red, almost red-brown, and is covered with irregular whitish or flesh-brown fragments of its universal veil. When young it is almost spherical, later convex; when mature it is flat with indistinct fine lines around the margin; 50-150mm (2-6in) in diameter.

The gills when young are white; when mature they become red or covered with reddish-brown flecks, 8-12mm (⅓-½in) wide, close together, not attached to the stipe and rounded near the cap margin.

Stipe Cylindrical, with a bulbous base; when young it is solid, later it develops a woolly filling and later becomes hollow; 60-180mm (2¼-7in) long, 15-40mm (⅝-1½in) thick, with a smooth or scaly surface; all white except its reddish base. In its upper third it has a thick white ring which is strikingly lined down the outside. The stipe is easily separated from the cap.

Flesh Thick, fleshy and brittle, off-white; when cut it turns pinkish; the pink tinge deepens at its base, particularly where it is damaged by insect larvae. When raw it is indigestible, tasting rather unpleasant — like raw potato — and it has little smell.

Spore deposit White.

Spores 8-9 × 5.5-7μm, elliptical; smooth; colourless.

Habitat It grows from June to November in mixed woodland, usually in clumps. It is more common in lowlands than in mountains.

Use This good, edible mushroom is especially suitable for frying. It is easily recognized by the way its flesh turns pink at the base, and by its thick grooved ring.

Remarks None of the poisonous sorts of amanita is likely to be mistaken for The Blusher. It differs from them because of the characteristic which gives it its popular name, turning pink when cut.

THE PANTHER FALSE BLUSHER, PANTHER CAP

Poisonous

Amanita pantherina (DC. ex Fr.) Krombh.

Cap When young it is dusky brown; the mature fungus is light brown to yellowish-brown, covered with fragments of its universal veil. When young it is almost spherical, later becoming hemispherical or convex and finally flat; its margin it distinctly grooved; 60-120mm (2¼-4¾in) diameter.

The gills are white, closely-packed, 6-10mm (¼-⅜in) wide, not attached to the stipe and rounded near the cap margin.

Stipe Cylindrical, thicker and bulbous at the base; 60-150mm (2¼-6in) long, 6-20mm (¼-¾in) thick; white, smooth and shiny. The upper section bears a smooth ring hanging down which usually soon disappears; above the bulb, wrapped in the volva which is attached to it, there are sometimes one to three rings of cracks in the surface of the stipe. When young it is solid; the mature stipe is hollow; it is easily separated from the cap.

Flesh Thin, fleshy, soft; it retains its colour when cut; it has a mild flavour and little smell.

Spore deposit White.

Spores 10-12 × 7-8μm, elliptical; smooth; colourless.

Habitat It grows from August to the end of October in mixed woodland; or singly in small groups. It is quite frequent in lowlands and is also found in mountains.

Use The Panther is a highly poisonous fungus which can be fatal.

Remarks It can be difficult to distinguish between The Panther and the larger *Amanita spissa* (Fr.) Opiz. The cap of *A. spissa* is grey or brownish-grey, thickly covered with off-white or greyish fragments of its universal veil; it has a smooth margin with no grooves. Its stipe is very stout with a bulbous base; on the upper section is a thick grooved ring. The flesh smells either earthy or like radish. This mushroom is not poisonous, but it is not advisable to pick it in case you confuse it with The Panther.

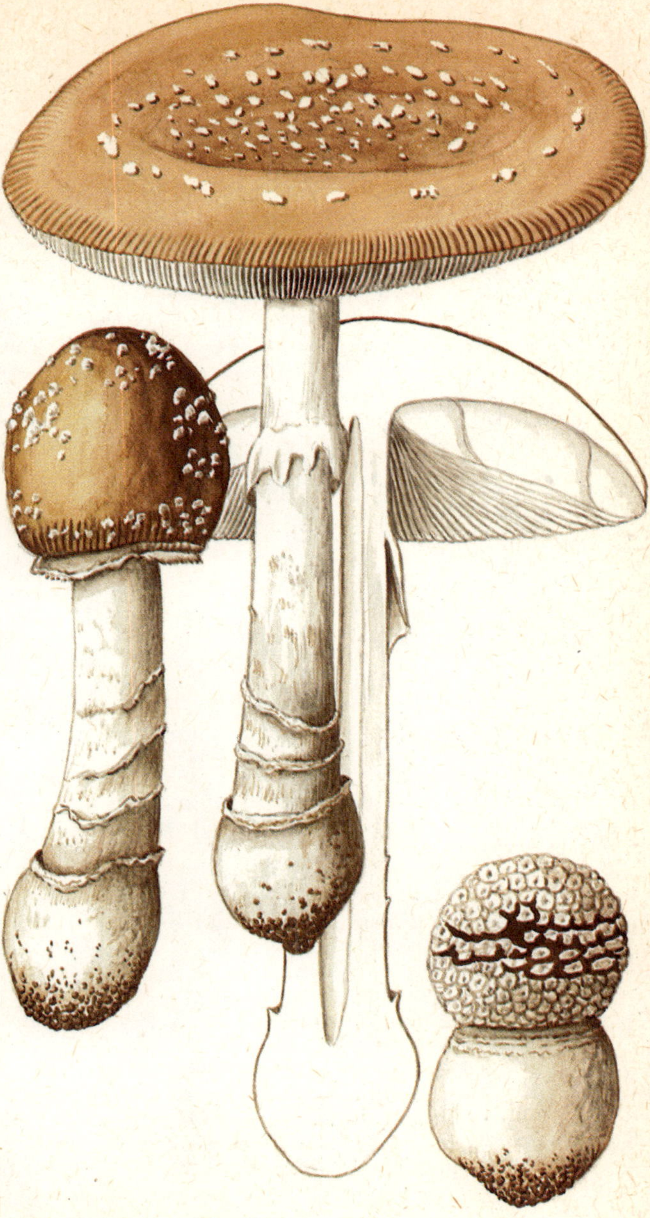

†

DEATH CAP **Fatal**

Amanita phalloides (Fr.) Link

Fruiting body In the early stage of its development the Death Cap is egg-shaped and completely enclosed in its white, membranous universal veil. After the veil breaks the cap and stipe appear and the rest of the veil remains at the base of the stipe as a free white cup-like volva.

Cap Very variably coloured, but almost always with a greenish tinge. It is always darkest in the centre; occasionally it is orange-yellow or white. When young it is bell-shaped or convex; the mature cap is flat, 60-150mm (2¼-6in) in diameter. The cuticle has a pearly sheen and is streaked with slender radiating fibrils; it is smooth and bare. The gills are white, closely packed, 8-12mm (⅓-½in) wide, not attached to the stipe and rounded near the cap margin.

Stipe Cylindrical, 50-150mm (2-6in) long, 10-25mm (⅜-1in) thick, with a large bulb at its base which is covered by the remains of the volva. When young it is solid; later it has a woolly filling. A mature stipe is hollow and easily separated from the cap. Its surface is usually white; rarely it is greenish-yellow or greyish with greenish stripes. In its upper section it has a thin, white ring hanging down.

Flesh Soft, pliable and white; under the cuticle of the cap it is brownish-yellow. It retains its colour when cut. The young fungus has little smell but a mature one smells like old raw potatoes.

Spore deposit White.

Spores 8-10μm, almost spherical; smooth; colourless.

Habitat It grows from July to October in deciduous woods — in rare cases also in coniferous woods — singly or in small groups. It is most common in warm lowland areas.

Use The Death Cap is the most dangerous poisonous fungus. Every year people die from eating it.

Remarks Another *Amanita* with a poison of similar strength is the Spring Amanita *Amanita verna* (Bull. ex Fr.) Pers. ex Vitt, which is very similar to a typical Death Cap. The only difference between them is that *A. verna* is white. It grows in deciduous woods in warm lowland regions from late spring to the end of summer.

DESTROYING ANGEL **Fatal**

Amanita virosa (Fr.) Bertillon

Cap It is usually white, yellowish or creamy at its centre; at first conical, later bell-shaped; when mature it is slightly convex with a smooth surface. In damp weather the young fungus is slightly sticky, otherwise it is dry and shiny; 60-100mm (2¼-4in) in diameter.

The gills are white, close together, 5-8mm (3/16-5/16in) wide, free, not attached to the stipe, near the cap margin they are rounded.

Stipe Cylindrical, 80-120mm (3⅛-4¾in) long, 8-15mm (5/16-⅝in) thick, thinner at the top; it has a bulbous base wrapped in a free white volva; when young it is solid, later becoming hollow. Its surface is white, silky with a sheen and fibrously scaly; the upper section bears a thin ring which hangs down.

Flesh White; retains its colour when cut; has a mild flavour and little smell.

Spore deposit White.

Spores 7-10μm, spherical; smooth; colourless.

Habitat It grows from August to October in coniferous woods in hilly and mountainous regions. It is very rare in the leafy woods of lower areas.

Use It is deadly poisonous.

Remarks The Destroying Angel may be confused with the Spring Amanita, *Amanita verna* (Bull. ex Fr.) Pers. ex Vitt., which is also white but looks very much like the Death Cap, *Amanita phalloides* (Fr.) Link. *A. verna* grows in the deciduous woodlands of warm, lowland regions from late spring to the end of summer.

FALSE DEATH CAP

Inedible

Amanita citrina (Schaeff) ex Roques
Syn.: *Amanita mappa* (Batsch ex Lasch) Quél.

Cap At first it is lemon or sulphur-yellow with a greenish tinge; later it is off-white and speckled with orange-yellow or brownish fragments of its universal veil, which often disappear completely. When young it is hemispherical; later it becomes convex and then flat; 50-100mm (2-4in) in diameter. In damp weather it is sticky and matt; in dry weather it has a sheen. The gills are closely packed, white, 6-12mm (¼-½in) wide, not attached to the stipe; near the cap margin they are rounded.

Stipe Cylindrical; when young it is solid, later hollow; 70-150mm (2¾-6in) long, 10-20mm (⅜-¾in) thick. Its lower part ends in a distinctive hemispherical bulb covered in a white volva which looks as if it has had its upper margin cut off. The surface of the stipe is off-white or greenish-yellow and smooth; the upper section bears a soft yellowish ring which hangs down.

Flesh White; retains its colour when sliced; tastes mild at first but later unpleasant; its smell is like that of old raw potatoes.

Spore deposit White.

Spores 7-10μm, almost spherical; smooth; colourless.

Habitat It grows from the end of August to the first half of November in mixed woodland. It is a very common autumn fungus.

Use Inedible but harmless. Formerly it was thought to be poisonous. At first sight it often looks like the Death Cap, *Amanita phalloides* (Fr.) Link, particularly when it loses the last remains of the universal veil from its cap.

Remarks A very similar fungus is *Amanita gemmata* (Fr.) Gill., which has an orange-yellow cap and a smaller fruiting body. It grows from May to October in mixed woodland. It is poisonous.

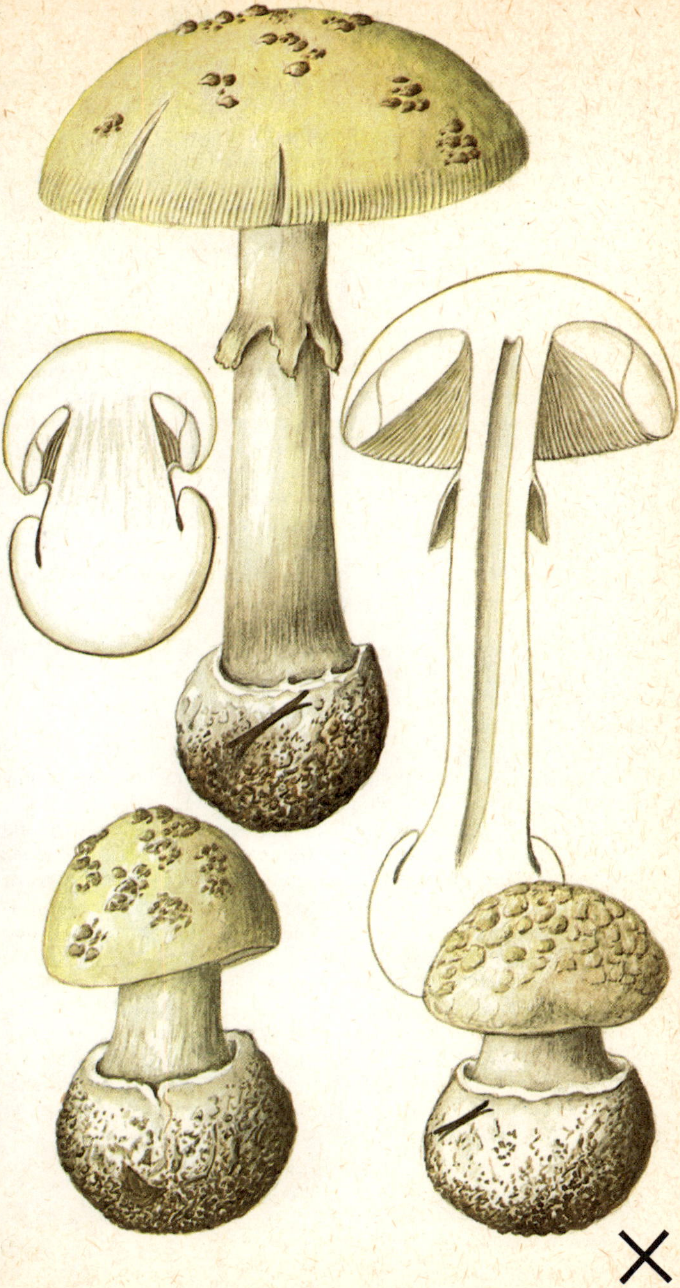

DEER MUSHROOM, FAWN AGARIC **Edible**

Pluteus cervinus (Schaeff. ex Fr.) Kumm.
Syn.: *Pluteus atricapillus* (Secr.) Sing.

Cap when young it is bell-shaped, later convex then almost flat, often with a hump; dark brown, grey-brown or grey; bare; when old it often develops cracks running from the margin to the centre of the cap, especially in dry weather; 50-120mm (2-4¾in) in diameter.

The gills are 10-15mm (⅜-⅝in) wide, closely packed and not attached to the stipe; near the cap margin they are rounded; white when young, later flesh-pink.

Stipe Cylindrical, solid, 70-120mm (2¾-4¾in) long, 10-15mm (⅜-⅝in) thick, white or greyish-white with a black fibrous surface extending down the stem; it is easily separated from the cap.

Flesh Soft and whitish; keeps its colour when sliced; it has a mild taste and smell.

Spore deposit Flesh-pink.

Spores 8-9 × 5-9μm, elliptical; smooth; pale pink.

Habitat It grows from April to the first half of December on the boles or stumps of deciduous trees; singly or in small clumps.

Use It is an edible but not very tasty mushroom. Its flesh tends to absorb water which attracts insects to lodge in the gills.

Remarks A very similar mushroom is *Pluteus atromarginatus* (Konr.) Kühn. It has a blackish cap, rather scaly at the centre, and gills which are black at the edges. It grows on the dead trunks of coniferous trees.

ROSE-GILLED GRISETTE **Edible**

Volvariella speciosa (Fr. ex Fr.) Sing.
Syn.: *Volvaria speciosa* (Fr.) Kumm.

Cap When young it is bell-shaped; later it becomes convex and then completely flat, often having a rounded hump at its centre; the surface is usually a dirty white or grey with a greyish-brown centre; smooth and bare; in damp weather it is sticky, at other times dry and with a silky sheen; 60-120mm (2¼-4¾in) in diameter.

The gills are 8-12mm (⅓-½in) wide, closely set, not attached to the stipe; rounded near the cap margin; white when young, then flesh-pink.

Stipe Cylindrical, solid, without a ring; 100-150mm (4-6in) long, 10-30mm (⅜-1⅛in) thick, with a bulbous base wrapped in a white or ochre free volva; its surface is white and smooth, with a silky sheen.

Flesh White; retains its colour when cut; hardly any flavour or smell.

Spore deposit Flesh-pink.

Spores 12-18 × 8-10μm, a compact ellipse; pale pink.

Habitat It grows in May and June in gardens and parks in well-manured soil, singly or in clumps.

Use It is an edible mushroom, but not very tasty. Formerly it was thought to be poisonous.

Remarks *Volvariella bombycina* (Schaeff. ex Fr.) is a similar mushroom, growing on live and dead trunks of deciduous trees. It is a beautiful mushroom, with a silky fibrous, white or light yellow cap. Its stipe is white, with a white or yellowish free volva at its base. It is edible. Although edible, with much the same taste as the Deer Mushroom, it is far rarer.

PARASOL MUSHROOM

Edible

Macrolepiota procera (Scop. ex Fr.) Sing.

Syn.: *Lepiota procera* (Scop. ex Fr.) Kumm.

Cap At first it is completely brown; during development the brown cuticle gradually cracks into rough, loose scales, among which fibrous white or creamy flesh can be seen. When young it is egg-shaped, later conical then arched; the mature cap is flat with a pronounced brown hump; the margin of the cap is usually a downy white. The gills are closely set and convex, 10-18mm (⅜-11/16in) wide, white when young, turning creamy.

Stipe Cylindrical, with a strikingly thick bulbous base; 200-400mm (8-16in) long, 20-40mm (¾-1½in) thick. The young stipe has a woolly interior; when mature it is hollow and easily separated from the cap. Its surface is covered with a continuous brown cuticle which becomes irregularly cracked and speckled with raised fragments. The upper section bears a double ring which can slide along the stipe like a collar. Its base is white and downy.

Flesh White, wooly in the cap, woody and fibrous in the stipe; it has a pleasant smell and a mild flavour.

Spore deposit White.

Spores 15-20×10-13μm, elliptical; smooth; colourless.

Habitat It grows from July to October in mixed woodland, mainly in warm, sunny, grassy clearings and glades; it may be solitary or in groups.

Use Only young, medium-sized caps of this excellent mushroom are edible. They are prepared like a Vienna cutlet. Before it is cooked the cuticle should be cut off because it cannot be peeled.

Remarks A similar mushroom is the Shaggy Parasol, *Macrolepiota rhacodes* (Vitt.) Sing., which has more closely-set scales on the cap, a bare stipe and reddish flesh. It is edible, but less tasty than *Macrolepiota procera*.

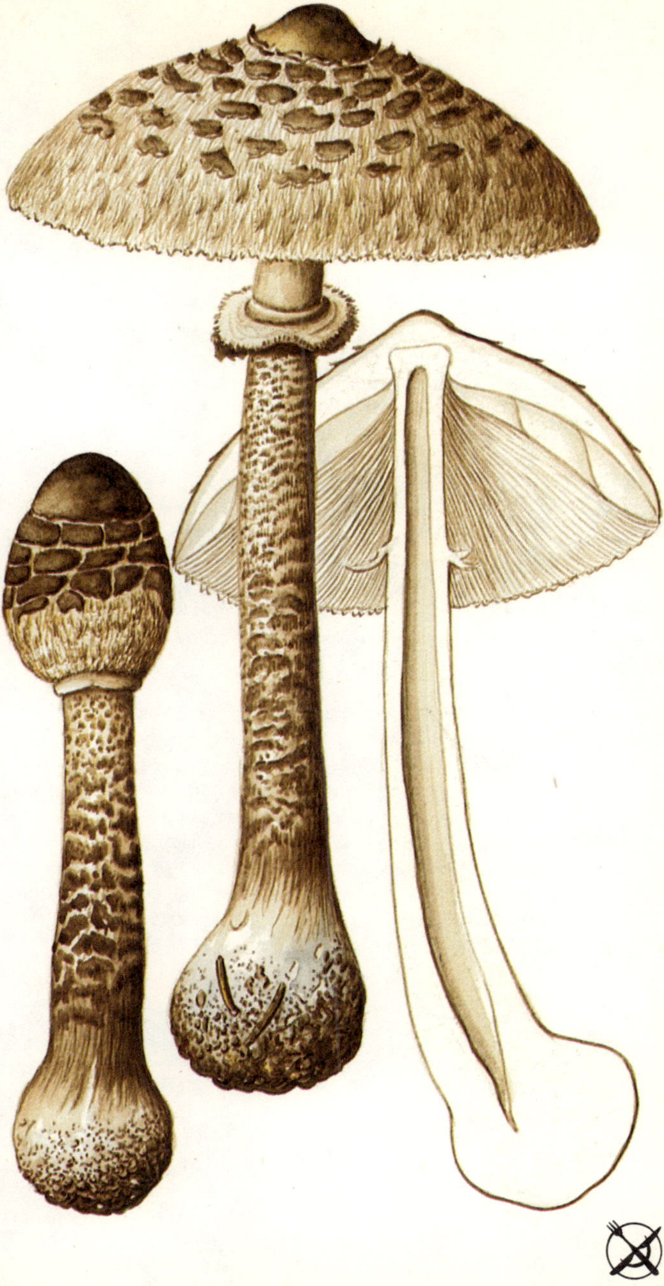

SHAGGY PARASOL **Edible**

Macrolepiota rhacodes (Vitt.) Sing.
Syn.: *Lepiota rhacodes* (Vitt.) Quél.

Cap When young it is almost spherical, but grows convex and flat, with a small hump in its centre; 50-150mm (2-6in) in diameter. The surface is cracked into rough, loose scales, at first pale brown, later dark brown. The hump is darker.

Its gills are at first white, but the mature gills are usually reddish at their edge, 12-15mm (½-⅝in) wide, close together, convex and not attached to the stipe.

Stipe Cylindrical, with a thickened bulbous base; when young it is off-white or creamy-orange-yellow, then brownish-yellow. When mature it is hollow and bare. The upper section has a soft, fibrous ring which can slide along the stipe.

Flesh White; turns orange-red when cut and later becomes brown; it has a mild flavour and a spicy smell.

Spore deposit White.

Spores 10-12 × 6-7μm, elliptical; smooth; colourless.

Habitat It grows from July to October in mixed woodland rich in humus; it is also common in acacia copses.

Use Even though it is not so full of flavour as the Parasol Mushroom, it is a good edible mushroom.

Remarks *Macrolepiota puellaris* (Fr.) Mos. is similar to the Shaggy Parasol. It has a paler cap and its flesh turns red only at the base of the stipe. It grows from the end of August to October in coniferous woods. It is a good edible mushroom.

HORSE MUSHROOM, HORSE AGARIC

Edible

Agaricus arvensis Schaeff. ex Fr.
Syn.: *Psalliota arvensis* (Schaeff. ex Fr.) Kumm.

Cap White or cream, smooth, bare, with a silky sheen; when bruised it turns yellow. It is egg-shaped when young but later becomes convex; the mature cap is flat; 50-150mm (2-6in) in diameter.

The gills are at first greyish-white, later greyish-brown; when mature they are dark brown, almost black; 8-12mm (⅓-½in) wide, closely packed and not attached to the stipe.

Stipe Cylindrical, thickened at the base, 60-150mm (2¼-6in) long, 10-25mm (⅜-1in) thick. When young it is usually solid, later it becomes hollow; it can easily be separated from the cap. Its surface is smooth, bare and white; when bruised it turns yellow. The upper section bears a large skirt-like ring.

Flesh Soft and pliable, white when young; very old mushrooms are sometimes yellowish. It has a mild flavour and smells like liquorice.

Spore deposit Very dark brown.

Spores $6\text{-}8 \times 4\text{-}5\mu m$, egg-shaped or elliptical; smooth; brownish-purple.

Habitat It grows from June to October in coniferous woods, rarely in deciduous woods, singly or in small groups. It is most common in spruce woods on mountains.

Use This is perhaps the tastiest of all agarics. It can be cooked and prepared in many different ways like cultivated mushrooms. It is a substantial fungus with a big, fleshy body.

Remarks It is easy to mistake the moderately poisonous Yellow Stainer, *Agaricus xanthodermus* Genev. for the Horse Mushroom. These two are very similar, especially in the early stages of their development. As a guide, remember that the Yellow Stainer when young has pink gills and the flesh of its stipe base is a striking bright yellow.

BROWN WOOD MUSHROOM **Edible**

Agaricus sylvaticus Schaeff. ex Krombh.
Syn.: *Psalliota sylvatica* (Schaeff. ex Krombh.) Kumm.

Cap At first it is pale orange-yellow but soon turns cinnamon or dark brown, streaked with fibrous scales; when young it is almost spherical, later convex; when mature it spreads flat; 40-90mm (1½-3½in) in diameter.

The gills are 5-8mm (3/16-5/16in) wide, closely set and not attached to the stipe; at first they are greyish-red, later brownish, when mature chocolate brown; they have a pink tinge, but it is not very distinct at any time.

Stipe Cylindrical, bulbous at the base; 50-120mm (2-4¾in) long, 10-25mm (⅜-1in) thick; solid when young. The mature stipe is hollow and easily separated from the cap. Its surface is smooth; minute fibrils are visible which are white or greyish; the upper section bears a wide collar-like ring. When bruised the stipe turns red.

Flesh White, but when cut it turns carmine red; it tastes good but has little smell.

Spore deposit Dark brown.

Spores 5.5-6×3.5-4μm, egg-shaped; smooth; reddish-brown.

Habitat It grows from July to November in coniferous woods, singly or in small clumps. It is most common in spruce woods.

Use This good edible mushroom is suitable for any kind of cooking, but only a few people gather it because of its unattractive appearance and the fact that its flesh turns red.

Remarks A similar mushroom is *Agaricus haemorrhoidarius* Kalchbr. et Schulz., but it grows bigger, has thicker scales at its cap and when young its gills are pink. It is found in spruce woods, usually on limey soil. It is more fleshy than the Brown Wood Mushroom.

FIELD MUSHROOM, PINK BOTTOM **Edible**

Agaricus campestris L. ex Fr.
Syn.: *Psalliota campestris* (L. ex Fr.) Kumm.

Cap White, with a silky sheen, bare, or partly covered with scattered greyish-brown scales. When young it is nearly spherical (the so-called 'button mushroom'), later it becomes convex then almost flat; 50-120mm (2-4¾in) in diameter. On young mushrooms the partial veil connects the margin of the cap with the stipe and covers the gills until it is torn.

The gills when young are pink, turning to dark brown or almost black when mature; closely set; 8-10mm (about ⅜in) wide, and not attached to the stipe. Cylindrical, solid, 40-80mm (1½-3⅛in) long, 20-40mm (¾-1½in) thick, easily separated from the cap. Its surface is smooth and bare; white except at the base which is often rusty-brown to deep brown. The upper section bears a white membranous loose ring.

Flesh Compact, pliable and white; when cut it may turn pink. It has a very pleasant flavour and smell.

Spore deposit Dark brown.

Spores $7\text{-}10 \times 5\text{-}6\mu m$, elliptical; smooth; reddish-brown.

Habitat It grows from May to November in meadows, pastures, stubble-fields, gardens, and wayside ditches. It is common on well-manured soil. It forms large clumps.

Use A good and tasty mushroom; used in soups, and egg and meat dishes, also as a stuffing for roast chicken.

Remarks Pavement Mushroom *Agaricus bitorquis* (Quél) Sacc. is very similar. Its cap is always bare, and there are two rings on the stipe. It grows from May to November, and is most common among ruins and by urban roadsides. It is edible.

SHAGGY CAP, LAWYER'S WIG **Edible**

Coprinus comatus (Müll. ex Fr.) S. F. Gray

Cap White, orange-yellow at the top; its surface is fibrous and scaly. At first it is cylindrical, later bell-shaped, 50-100mm (2-4in) tall, 30-60mm (1⅛-2¼in) wide. The mature cap develops ray-like cracks from its margin which turns up. The whole cap, with its gills, gradually blackens and liquifies.

The gills are at first white, turning pink to violet and finally black; 10-18mm (⅜-11/16in) wide, closely set and slender. When mature they turn liquid.

Stipe Cylindrical, hollow, 100-200mm (4-8in) long, 8-20mm (5/16-¾in) thick, easily separated from the cap. Its surface is smooth and white; it has a loose ring which often drops down to the base.

Flesh White and juicy; the cap is thin and brittle; the stipe is soft and fibrous. It has a mild flavour and a pleasant but faint smell.

Spore deposit Black.

Spores 10-14 × 6-8μm, elliptical; smooth; black.

Habitat It grows from April to November in clumps. It is most common in gardens, parks and meadows, beside woodland paths and along urban grass verges. It is easily seen from a distance.

Use While very young it is edible and tasty. It is used for soups and stews. Once the gills begin to turn pink the mushroom is no longer fit to eat.

Remarks A similar fungus is the Grey Ink Cap *Coprinus atramentarius* (Bull. ex Fr.) Loud. Its cap is grey or greyish-brown and grooved at the margin. Its stipe is white, with a faint ring. It grows in gardens, on compost, in ruins and in meadows and woods, in large clumps from April to November. The young mushrooms are edible, but if taken in combination with alcohol can cause nausea.

SULPHUR TUFT

Poisonous

Hypholoma fasciculare (Huds. ex Fr.) Kumm.
Syn.: *Namatoloma fasciculare* (Huds. ex Fr.) P. Karst.

Cap Sulphur yellow, usually orange or tawny in the centre. When young it is hemispherical, with a smooth, bare surface; 30-60mm (1⅛-2¼in) in diameter. The gills are at first sulphur yellow, becoming green and when mature almost black, 4-6mm (up to ¼in) wide; close together, rounded near the stipe and attached to it. When young they are usually covered by a thin membranous veil which soon tears.

Stipe Irregularly cylindrical, often curved to some extent, hollow, 50-100mm (2-4in) long, 3-7mm (⅛-¼in) thick, with a smooth, bare sulphur-yellow surface. When young it has a faint ring at the top, formed of fragments of veil; this ring soon disappears.

Flesh Thin, pliable, sulphur yellow; it has a bitter taste and an unpleasant, earthy smell.

Spore deposit Deep purple.

Spores 5.5-7×4-5μm, elliptical; smooth; at first they are purple, later brown with a purple tinge.

Habitat It grows freely from April to December on dead tree boles, forming large clumps.

Use It used to be considered inedible merely because of its bitter taste. The latest research has proved that it is poisonous.

Remarks The Smoky-gilled Woodlover *Hypholoma Capnoides* (Fr. ex Fr.) Kumm., is similar. Its gills are at first whitish, turning greyish-blue. It grows on the boles of coniferous trees only. Also similar is the Brick-red Agaric *Hypholoma sublateritium* (Fr.) Quél. Its cap is reddish-yellow to brick-red; its gills are yellowish turning yellowish-brown and eventually dark brown to black. It grows from September to November on the boles of coniferous and deciduous trees. It is inedible.

†

RED-STAINING INOCYBE **Fatal**

Inocybe patouillardii Bres.

Cap When young it is white, later pale orange-yellow or reddish-yellow; when mature it is brownish-red, silky and fibrous. At first it is a bell-shaped cone, later flat with a hump in its centre, usually with radiating cracks; often with a turned up margin; 25-90mm (1-3½in) in diameter.

The gills are 5-8mm (3/16-5/16in) wide, moderately close together; notched where they join the stipe; white, later olive-brown; when mature the cap has reddish-brown spots; white and scaly at the edge.

Stipe Cylindrical, slightly bulbous at its base; solid, 30-120mm (1⅛-4¾in) long, 8-20mm (5/16-¾in) thick; white, later with red spots or almost all red, with longitudinal fibrils showing as fine lines.

Flesh White; when cut it goes partly red; it has a mild flavour and a fruity smell.

Spore deposit Yellowish-brown.

Spores 9-14 $\times$ 5-8μm, kidney-shaped; smooth; pale orange-yellow.

Habitat It grows from the end of May to the beginning of October in deciduous woods and parks, usually under beeches, singly and scattered.

Use This highly poisonous fungus can cause death in a few minutes when the heart stops or the patient suffocates. A mild dose may cause poisoning which lasts about 12 hours; a further 12 hours is necessary for full recovery.

Remarks There are approximately 130 species of *Inocybe*. Though a few of them are edible they are not recommended because it is easy to mistake a poisonous *Inocybe* for an edible one.

INOCYBE NAPIPES

Poisonous

Inocybe napipes Lange

Cap Dark brown, deeper in the centre; at first a bell-shaped cone, later flat and extended with a fairly large hump in the centre; when young it is usually bare, later minutely fibrous; when mature it is markedly fibrous with radiating cracks; 30-60mm (1⅛-2¼in) in diameter.

The gills are at first off-white then pale grey; mature gills are pale brown, 4-6mm (about ¼in) wide; close together, at first attached to the stipe then detached.

Stipe Cylindrical, thinner at the upper end, bulbous at the base, solid; 50-80mm (2-3⅛in) long, 4-8mm (⅙-⅓in) thick, with a fine, longitudinally fibrous surface, a paler shade than the cap.

Flesh White or pale-cream with a brown tinge in the stipe (except the bulbous base). It has a mild taste and is virtually odourless.

Spore deposit Pale orange-yellow.

Spores 9-10 × 5-6μm, egg-shaped, with 5-6 irregular projections; pale ochre.

Habitat It grows from August to the end of October in damp grassy places in deciduous woods; singly or in small clumps, usually under birches.

Use This fungus is poisonous.

Remarks The Conical Inocybe *Inocybe fastigiata* (Schaeff. ex Fr.) Quél, is dangerously poisonous. It has a pale orange-yellow cap, off-white stipe and white flesh. It grows from June to October in mixed woodland. It is much more frequent than *Inocybe napipes.*

†

CORTINARIUS ORELLANUS **Fatal**

Cortinarius orellanus Fr.

Cap Cinnamon brown or brownish-red, usually dark in the centre, paler at the margin; finely fibrous or somewhat scaly, dry and smooth. When young it is usually hemispherical or bell-shaped, later convex, with a wavy margin, bent downwards, and with a rounded hump in its centre; 30-80mm (⅛-3⅛in) in diameter.

The gills are 4-8mm (⅛-⅓in) wide, convex, widely-spaced and rather thick; notched where they join the stipe; at first pale orange, when mature rust-brown. The young gills are covered with a creamy-yellow gossamer veil.

Stipe Cylindrical, thinner at the base, solid, 40-80mm (1½-3in) long, 5-15mm (3/16-⅝in) thick, with yellowish surface; the lower part goes rusty-brown; it has a slight sheen. The remains of the fine veil in the upper section soon disappear.

Flesh Yellowish; slightly rusty-red to rusty-brown in parts. It tastes slightly sour and has little smell.

Spore deposit Rust-brown.

Spores 9-12 $\times$ 5.5-7μm, elliptical; covered with tiny warts; ochre-brown.

Habitat It grows from July to October in deciduous woods. It is most common in sandy soil under oaks and birches, singly or in small clumps.

Use Poisoning by this fungus often proves fatal.

Remarks *Cortinarius cinnamomeoluteus* P. D. Orton, is similar. It has paler cap and gills and its flesh smells like radishes. It grows from July to October in mixed woodland. It is inedible but not poisonous.

THE GYPSY, GYPSY MUSHROOM **Edible**

Rozites caperata (Pers. ex Fr.) P. Karst.
Syn.: *Pholiota caperata* (Pers. ex Fr.) Kumm.

Cap At first it is pale apricot with a light blue or somewhat purple bloomed surface; when that disappears the cap is sandy. When young it is egg-shaped, later convex; when mature it spreads out flat; in dry weather it has radiating wrinkles and cracks around the margin; 40-100mm (1½-4in) in diameter.

The gills when young are a dull yellow, later rusty-brown; 4-8mm (1/6-⅓in) wide, closely set; notched where they join the stipe; at first covered with a white membranous partial veil.

Stipe Cylindrical, often swollen at the base and solid; 50-120mm (2-4¾in) long, 10-25mm (⅜-1in) thick, with a silky fibrous surface, flaky or minutely scaly; yellowish above the ring, pale orange-yellow under the ring, with fragments of purplish universal veil. The ring is usually membranous; at first white, later yellowish, loose and permanent.

Flesh White or pale cream; under the cuticle of the cap and stipe it is yellowish. It has a mild flavour and is virtually odourless.

Spore deposit Rust-brown.

Spores 11-13 × 7.5-9μm, almond-shaped with a warty surface; ochre.

Habitat It grows from the end of July to October, almost always in large groups in spruce woods. It is rarely found in deciduous woods.

Use This good edible mushroom can be prepared in many ways. It is very good in meat dishes.

Remarks No poisonous fungus can be mistaken for *Rozites caperata*. A few species of *Cortinarius* could be taken for it, but even this is unlikely.

CHANGEABLE AGARIC **Edible**

Kuehneromyces mutabilis (Fr.) Sing. et A. H. Smith
Syn.: *Pholiota mutabilis* (Fr.) Kumm.

Cap In damp weather it is reddish-brown or cinnamon brown and sticky; in dry weather it gradually dries from the centre to the margin, so that the centre is usually pale orange-yellow and the margin brownish, with fine radiating lines of gills seen through its thin flesh. When young it is hemispherical, later convex; when mature it becomes completely flat, smooth and bare, with a rounded hump in the centre; 30-60mm (1⅛-2¼in) in diameter.

The gills are at first sandy, later rusty-brown, 3-5mm (⅛-3/16in) wide; closely set; fully attached or running a little way down the stipe; when young they are covered with a whitish membranous veil.

Stipe Cylindrical and hollow; 60-120mm (1¼-4¾in) long, 5-10mm (3/16-⅜in) thick; above the ring it is brownish-yellow and bare, under the ring it is brown and scaly, becoming dark brown at the base.

Flesh In the cap it is thin, whitish, soft and watery; in the stipe it is usually hard and rusty brown. It has a mild flavour and a pleasant smell.

Spore deposit Rusty brown.

Spores 6-7×4-4.5μm, shaped like an egg with its top cut off; smooth; light brown.

Habitat It grows from April to December on the boles of dead deciduous trees, usually in large clumps. It is a very common mushroom, Although it begins to grow in early spring it is most abundant in autumn.

Use This good edible mushroom is suitable for soups, sauces and meat dishes. Only the caps are eaten because the stipe is tough.

Remarks A very similar fungus is *Galerina marginata* (Fr.) Kühn. *(Galerina unicolor)*, which grows in large clumps on the trunks of coniferous trees in autumn. However, its stipe under the ring is smooth not scaly and it never grows on deciduous trees. It is not edible.

SWEETBREAD MUSHROOM, THE MILLER, PLUM AGARIC

Edible

Clitopilus prunulus (Scop. ex Fr.) Kumm.

Cap White or greyish-white; at first with a fine felt-like surface, later smooth and shiny; in damp weather slightly sticky. When young it is hemispherical, later cushion-shaped with a down-curved margin. It is usually flat and sunken in the centre when mature; 30-100mm (1⅛-4in) in diameter.

The gills are at first white, turning flesh-pink, 2-4mm (about ⅛in) wide, closely set and attached well down the stipe.

Stipe Cylindrical, thinning towards the base; solid, 30-60mm (1⅛-2¼in) long, 7-20mm (¼-¾in) thick, sometimes off-centre; white or pale grey. Its top looks as if sprinkled with flour; the base is white with a felted appearance, often covered with leaf-like structures.

Flesh Soft and flexible; white, retaining its colour when cut; it has a mild, pleasant taste and smells like freshly-ground flour.

Spore deposit Pinkish.

Spores 8-14 × 5-6μm, spindle-shaped, with six descending ribs; light yellow with a light pink tinge.

Habitat It grows from July to November in mixed woodland, almost always in great abundance.

Use This good edible mushroom can be used in soups and meat dishes. It is also suitable for drying but is too soft for pickling.

Remarks *Clitopilus prunulus* may be confused with some of the poisonous sorts of *Clitocybe* which are also white. Bear in mind that the gills and spore powder of *Clitocybe* spp. are white, while the mature *C. prunulus* has flesh-pink gills and pinkish spore deposit.

LIVID ENTOLOMA

Poisonous

Entoloma sinuatum (Pers. ex Fr.) Kumm.

Syn.: *Entoloma lividum* (Bull. ex St-Am) Quél., *Rhodophyllus sinuatus*

Cap Whitish or greyish-yellow; when mature it is greyish-brown; smooth, bare, finely fibrous, with a sheen. When young it is hemispherical or a bell-shaped cone with a downward curved margin; later it becomes convex and, when mature, flat, sometimes with a rounded fleshy hump in the centre; 60-200mm (2¼-8in) in diameter.

The gills are 8-15mm (5/16-5/8in) wide; moderately close together; notched where they join the stipe; at first off-white, later creamy, then a yellowish salmon-pink; when mature they are flesh-coloured.

Stipe Cylindrical, sometimes thickened at the base. When young it is solid; when mature it either has a spongy filling or is hollow; 40-180mm (1½-7in) long, 8-30mm (5/16-1⅛in) thick, with the surface at first white, later orange-yellow; finely fibrous and grooved with a silky sheen.

Flesh White and shiny, retaining its colour when cut. It has a mild flavour and smells like freshly-ground flour.

Spore deposit Flesh coloured.

Spores 9-11 × 8-9μm, irregularly hexagonal; smooth; pale yellowish-pink.

Habitat It grows from the end of May to the beginning of October in airy, open deciduous woods, usually in clumps under beeches and oaks.

Use This is a highly poisonous fungus, and if much is eaten it may cause death. Other species of *Entoloma* are also poisonous, so it is better not to pick any of these fungi.

Remarks *Entoloma sinuatum* is very attractive and is often mistaken for the harmless Buckler Agaric *Entoloma clypeatum* (L. ex Hook.) Kumm. Sometimes it is very difficult to distinguish between them. *Entoloma clypeatum* has a cap that changes colour when damp and grows in gardens and meadows under bushes, while *Entoloma sinuatum* does not change colour when damp and grows in deciduous woods.

†

BROWN ROLL-RIM **Poisonous**

Paxillus involutus (Batsch ex Fr.) Fr.

Cap Yellowish-brown or olive-brown and smooth; at first with a fine felted appearance, later smooth, dry and matt; in damp weather very slimy. When young it is convex with the margin considerably down-curved to form a rim; later it becomes flat, often depressed in the centre; 40-120mm (1½-4¾in) in diameter. The gills are at first pale yellow, later olive-yellow and when mature they are rust-brown; 4-6mm (1/6-¼in) wide, closely set, extending down the stipe; when young they are usually wavy and may be reticulated near the stipe.

Stipe Cylindrical, solid, 30-60mm (1⅛-2¼in) long, 10-20mm (⅜-¾in) thick, orange-yellow to brownish-yellow, smooth and bare; turns brown when bruised.

Flesh Soft, flexible and juicy; at first it is pale yellow, later brownish-yellow; its base is a rusty-brown; it has a sour taste and little smell.

Spore deposit Yellowish-brown.

Spores 8-10 × 5-6μm, elliptical; yellowish-brown.

Habitat It grows freely from June to December in mixed woodland, almost always in large clumps.

Use It is a highly poisonous fungus, causing severe illness and death.

Remarks At first sight it is similar to some inedible species of *Lactarius*, which when scratched can be identified by oozing milky droplets.

†

GOMPHIDIUS GLUTINOSUS **Edible**

Gomphidius glutinosus (Schaeff. ex Fr.) Fr.

Cap At first it is bluish-grey to brownish-purple, later it turns paler to orange-yellow. When old it often has black spots. When young it is hemispherical or a blunt cone, then becomes convex. It is flat when mature, often depressed in the centre. The cuticle is smooth and covered with a thick layer of slime; after the slime dries the cap remains shiny.

When young the gills are white or grey, later dark grey, even black due to the ripening spores; they are usually 3-6mm (1/8-1/4 in) wide, very widely spaced and thick, attached well down the stipe.

Stipe Cylindrical, solid, 50-100mm (2-4in) long, 15-20mm (5/8-3/4 in) thick, with a very slimy surface; off-white near the top, shading to lemon-yellow or deep yellow lower down. When young the margin of the cap is connected to the stipe by a white, membranous veil which, after tearing, leaves a ring in the upper section; this ring soon disappears.

Flesh Soft and juicy; in the cap at first white, then greyish; in the bottom part of the stipe it is a deep yellow. The flavour is mild, almost sweet, and there is little smell.

Spore powder Almost black.

Spores 17-23 × 5-6.5 μm, cylindrical or spindle-shaped; smooth; brownish-grey.

Habitat It grows from July to November in woodland. It is most common in spruce woods, when it may grow singly or, more usually, in groups.

Use This tasty, edible mushroom is suitable mainly for soups and meat dishes. Before cooking it you need to peel off the cuticle of the cap and scrape the slime off the stipe. It is best to clean this fungus as soon as you find it, to prevent it from soiling other mushrooms already picked.

Remarks No poisonous fungus may be mistaken for *Gomphidius glutinosus*; you might mistake it for other kinds of *Gomphidius*, but all are edible.

GOMPHIDIUS RUTILUS **Edible**

Gomphidius rutilus (Schaeff ex Fr.) Lund. et Nannf.

Cap Yellowish-brown or wine-red, often with a greyish-purple tinge. When young it is a blunt cone with a considerably down-curved margin; later it becomes convex and when mature, it is flat with a fleshy hump in the centre; 30-100mm (1⅛-4in) in diameter. The cuticle is smooth, bare, sticky and slimy; when dry it has a sheen.

The gills are at first purplish-brown; as the spores ripen the gills become greyish-brown or very dark brown; 3-6mm (about ¼in) wide, very widely-spaced and thick, deeply attached down the stipe.

Stipe Cylindrical, usually thinner at the base, 50-100mm (2-4in) long, 10-15mm (⅜-⅝in) thick. Its surface is yellowish- or reddish-brown with longitudinal fibrils; around the upper section is a dry purple ring formed of fragments of partial veil; this ring soon disappears.

Flesh Soft and juicy; in the cap and the upper part of the stipe it is pale yellow or yellowish-red; in the lower part a rich pinkish-yellow; it has a mild, pleasant flavour and little smell.

Spore deposit Olive-black.

Spores $17\text{-}20 \times 6\text{-}8\mu m$, cylindrical, spindle-shaped, smooth; brownish-yellow.

Habitat It grows from July to November in coniferous woods, almost always in large clumps. It is most abundant in pine woods.

Use It is a good edible mushroom, suitable for soups, sauces, meat and egg dishes and for pickling. Unfortunately most people are not aware of the good flavour of this mushroom.

Remarks It is very similar to *Gomphidius helveticus*, which has a scaly, fibrous cap. It grows in mountain regions, especially under cedars.

INDIGO BOLETUS **Edible**

Gyroporus cyanescens (Bull. ex Fr.) Quél.
Syn.: *Boletus cyanescens*

Cap Creamy, straw-yellow or brownish-yellow, densely fibrous or shaggy; turns blue when bruised. When young it is hemispherical, later it is convex, becoming cushion-like when mature; usually symmetrical, but sometimes lop-sided; 35-120mm (1⅜-4¾in) in diameter.

The tubes are white or creamy, turning blue when cut, at first attached to the stipe, later detached; 3.5-8mm (½-⅓in) long. Pores are minute, at first whitish, later creamy; at maturity they are yellowish, turning blue when bruised.

Stipe Irregularly cylindrical, sometimes pot-bellied, often with a pointed base, symmetrical or off-centre; 50-100mm (2-4in) long, 10-30mm (⅜-1⅛in) thick, with a woolly filling when young; later hollow with the interior divided into many chambers. Its surface is pale yellow to yellowish-brown; bare under the cap, elsewhere fibrous, turning a deep blue when bruised or exposed to the air.

Flesh White or creamy and very brittle; immediately turning a deep blue when cut; the blue fades after a while. It has a mild flavour and is almost odourless.

Spore deposit Pale yellow.

Spore 8-15 × 4-6μm, elliptical; smooth; pale yellow.

Habitat It grows from July to October in leafy deciduous lowland woods where it is usually found under oaks in sandy soil. It may be found alone or in small groups.

Use This edible and tasty mushroom is suitable for sauces, in spite of its blue colour, and for drying.

Remarks No poisonous fungus can be mistaken for the Indigo Boletus, but in very rare cases it could be confused with *Boletus junquilleus* (Quél.) Boud., whose flesh also turns blue when cut.

CHESTNUT BOLETUS **Edible**

Gyroporus castaneus (Bull. ex Fr.) Quél.

Cap Rust to chestnut-brown; when young it is usually convex; when mature it is flat or cushion-shaped; 40-110mm (1½-4¼in) in diameter. The cuticle is at first velvety, becoming smooth. In dry weather it often develops a crazed surface.

The tubes when young are white; when mature they are yellow and do not turn blue when cut; at first they are attached to the stipe, later they become detached; almost 8mm (5/16in) long. Pores are minute and rounded, at first white, later yellow; when bruised they develop brown spots.

Stipe 35-80mm (1⅜-3⅛in) long, 8-30mm (5/16-1⅛in) thick, symmetrical or lop-sided, irregularly cylindrical or club-shaped; more or less deeply depressed, with a dry reddish-brown surface. When young it is solid, developing a woolly filling; when mature it is hollow, sometimes with many small interior chambers.

Flesh White; keeps its colour when cut; firm and fleshy when young, brittle when mature; has a mild flavour and little smell.

Spore deposit Pale yellow.

Spores 7-10 × 4-6μm, elliptical; smooth; almost colourless or very pale yellow.

Habitat It grows from July to November in mixed woodland and is most common on sandy soil in dry, warm regions. Scattered specimens grow singly.

Use It is an edible mushroom, but not so good as the Indigo Boletus. It becomes bitter when cooked, but is most suitable for drying, when it loses the bitter taste.

Remarks No mushroom can be mistaken for *Gyroporus castaneus*.

LARCH BOLETUS **Edible**

Suillus grevillei (Klotzsch) Sing.

Cap Lemon-yellow to orange-yellow, smooth and covered with slime. When young it is bluntly conical, later convex and at maturity it is flat, broad and cushion-like; 40-150mm (1½-6in) in diameter. The cuticle is rather difficult to peel off.

The tubes are at first yellow, turning more olive, attached to the stipe and 6-12mm (¼-½in) long. The pores are at first rounded, becoming angular; yellow when young, maturing to olive-yellow or brown-yellow; they turn pinkish-brown when bruised. There is a partial veil at first which is yellow near the stipe, yellowish-brown elsewhere. The outside is slightly slimy. In fine weather the veil dries quickly, leaving no ring.

Stipe Cylindrical or club-shaped, solid; 45-100mm (1¾-4in) long, 10-25mm (⅜-1in) thick with a ring in its upper section. Above the ring it is lemon or golden-yellow, often with a net-like pattern visible; under the ring it is usually yellowish-brown with fibres running down the stipe.

Flesh Yellow, sometimes partly pink; soft and juicy in the cap, hard and fibrous in the stipe, especially when mature; it has a mild flavour and little smell.

Spore deposit Olive-yellow.

Spores $7\text{-}10 \times 3\text{-}4\mu m$, cylindrical, elliptical and spindle-shaped; smooth; pale yellow.

Habitat Grows from June to November under spruces, mostly in mountain regions, almost always in large groups.

Use This good, edible mushroom can be prepared in many ways, but unfortunately it is often attacked by insect larvae.

Remarks A very similar mushroom is *Suillus nueschii* Sing, which also grows under spruces in high mountain regions. It differs by having pale grey pores and flesh which turns blue. It is edible.

SLIPPERY JACK, RINGED BOLETUS **Edible**

Suillus luteus (L. ex Fr.) S. F. Gray

Cap Chocolate-brown, tawny, sometimes reddish- or greyish-brown with a tinge of purple. When young it is hemispherical, later convex; when mature it is cushion-like, often with a rounded fleshy hump in the centre, 40-120mm (1½-4¾in) in diameter. Radiating fibrils lie at the surface of the cuticle which is very slimy and can be easily peeled.

The tubes are pale yellow when young, later dark yellow; attached to the stipe; 6-14mm (¼-9/16in) long, easily separated from the rest of the cap. Pores are minute, pale yellow when young, later bright yellow; when mature they are brownish-yellow.

Stipe Cylindrical, solid, 35-110mm (1⅜-4¼in) long, 10-25mm (⅜-1in) thick, lemon-yellow at the top, brownish and longitudinally fibrous toward the base. A white membranous veil, which at first connects the stipe with the margin of the cap, leaves its remains on the stipe as a dark brown or purple ring. Above the ring the stipe is stippled with yellowish-brown grains.

Flesh Soft and juicy in the cap and slightly fibrous in the stipe; at first it is off-white, later lemon-yellow; at the base of the stipe it is rusty-brown.

Spore deposit Brown.

Spores 7-10 × 3-3.5μm, elliptical and spindle-shaped; smooth; pale yellow.

Habitat It grows from the end of May to December in pine woods, usually in large colonies but sometimes singly. This is the most common, tasty and popular species of *Suillus*.

Use Slippery Jack is full of flavour and easily digested when used in soups, sauces and meat dishes. It can be pickled with other species of mushrooms but contains too much water to be dried successfully.

Remarks *Suillus fluryi* Huijsman is very similar at first sight, but unlike Slippery Jack it has no ring on its stipe.

SUILLUS PLACIDUS **Edible**

Suillus placidus (Bon.) Sing.

Cap At first off-white, pale yellow at the margin; when mature it is often tinged with purple. When young it is convex but becomes flat, sometimes with a slight depression in the centre; 50-120mm (2-4¾in) in diameter. The cuticle is smooth, bare and slightly slimy.

The tubes are 5-11mm (3/16-7/16in) long, attached to the stipe; when immature they are yellowish-white, changing through lemon-yellow to olive-yellow. Pores are minute, at first they are the same colour as the tubes; when mature they are olive-brown. They secrete milky droplets which often leave purplish-brown flecks when they dry.

Stipe Cylindrical, sometimes spindle-shaped at its base, solid; 30-80mm (1⅛-3⅛in) long, 7-20mm (3/10-¾in) thick, without a ring; white, with the top under the cap slightly yellow; when mature it has red or brownish-red spots running along its length.

Flesh White; above the tubes clear yellow, under the cuticle of the cap sometimes slightly purple; soft and juicy; it has a mild flavour and smell.

Spore deposit Orange-yellow or brownish-yellow.

Spores $7\text{-}11 \times 3\text{-}4\mu m$, elliptical and spindle-shaped; smooth; olive-yellow.

Habitat It grows from June to November under pines and cedars, usually in small groups. It is quite rare.

Use Although it is edible, it cannot be considered first class because its flesh quickly deteriorates. Eat only young mushrooms as soon as possible after picking.

Remarks *Suillus placidus* cannot be confused with any other mushroom. The nearest to it is *Suillus plorans* (Roll.) O. Kuntze which has a dark red cap and an orange-yellow spotted stipe without any ring. It grows in summer and autumn in mountainous cedar forests. It is edible but very rare.

SUILLUS GRANULATUS **Edible**

Suillus granulatus (L. ex Fr.) O. Kuntze

Cap At first reddish-brown or rust; when mature it is orange-yellow, bare and slimy; the cuticle is easily peeled off. When young it is usually hemispherical, becoming convex and finally cushion-like; 50-100mm (2-4in) in diameter.

The tubes are at first pale yellow, later orange-yellow; when mature they are brownish-yellow; 5-10mm (3/16-3/8in) long, attached to the stipe. Pores are of the same colour as the tubes; they are minute when young, but grow to 1mm (1/32in) in diameter. They secrete white milky drops which leave tiny brown flecks in the pores and at the top of the stipe when they dry.

Stipe Cylindrical, solid, without a ring; 40-60mm (1½-2¼in) long, 8-15mm (5/16-5/8in) thick; pale yellow on the surface with brown dots near the cap. When mature it is usually brownish towards the base.

Flesh In the young mushroom it is white, later turning yellow. The colour is most intense above the tubes and in the stipe. It retains its colour when cut. In the cap it is as soft as butter at first, turning watery when mature; in the stipe it is fibrous; the flavour is mild and pleasant, and there is little smell.

Spore deposit Yellowish-brown.

Spores $8\text{-}10 \times 2.5\text{-}4\mu m$, elliptical and spindle-shaped; smooth; pale yellow.

Habitat It grows from June to November in coniferous woods; alone or in groups in grassy clearings and along forest paths. It is most common under pine trees.

Use It is an excellent, edible mushroom which can be cooked in any way. It is widely used in soups, sauces and stews.

Remarks It is very similar to *Suillus fluryi* Huijsman, which differs from it by its darker cap. It is closely related to *Suillus luteus* and *Suillus granulatus*.

COW BOLETUS **Edible**

Suillus bovinus (L. ex Fr.) O. Kuntze

Cap Yellowish-brown to reddish-yellow; smooth, bare; very slimy in damp weather. When young it is convex, later flat, rarely depressed in its centre; 40-110mm (1½-4¼in) in diameter. The cuticle can be peeled away only in small pieces.

The tubes are at first greyish-yellow, later yellow, turning brown or olive-brown; 6-10mm (¼-⅜in) long, usually attached to the stipe — occasionally running a little way down the stipe; hard to separate from the cap flesh. Pores are at first randomly interconnected, later they turn toward the margin, long and angular; the same colour as the tubes.

Stipe Cylindrical, solid, sometimes thickened at the base; 30-100mm (1⅛-4in) long, 5-20mm (3/16-¾in) thick, with no ring; yellowish to rusty yellow at the top, with longitudinal slender fibrils; reddish-brown at the base.

Flesh Pliable and pale yellow in the cap; fibrous and reddish-brown in the stipe; the flavour is mild or slightly bitter; it has little smell.

Spore deposit Pale brown.

Spores 6-10 × 3-4μm, elliptical and spindle-shaped; smooth; pale yellow.

Habitat It grows from June to November in pine woods, usually in sandy soils. It appears in great profusion after heavy rain.

Use It is an edible mushroom but not so good as other sorts of *Suillus*. It is most suitable for pickling with more tasty mushrooms.

Remarks At first sight *Chalciporus piperatus* (Bull. ex Fr.) Bataille appears very similar, but differs in being smaller and having red pores and an acrid taste. It grows from July to November in mixed woodland. It is used as spicy addition to a variety of dishes.

VARIEGATED BOLETUS **Edible**

Suillus variegatus (Sw. ex Fr.) O. Kuntze

Cap The young mushroom is hemispherical with a down-curving margin; when mature it is cushion-shaped; 50-140mm (2-5½in) in diameter. Its surface is covered with olive or greyish-orange felt which gradually tears into minute scales that vanish when the cap ripens. When young its cuticle is greyish-yellow or greyish-orange, later brownish-red; when mature it is pale orange-yellow; slightly sticky and not easy to peel off.

The tubes are 8-12mm (5/16-½in) long, at first attached to the stipe, later slightly detached; yellow or pale orange, becoming dark olive when mature; turning slightly blue when bruised. The pores are at first minute, greyish-yellow; later they grow bigger and pale orange; when mature they are olive-brown, turning blue when bruised.

Stipe At first it is pot-bellied, later cylindrical or club-shaped, solid; 30-90mm (1⅛-3½in) long, 20-35mm (¾-1⅜in) thick, with a smooth surface of pale lemon-yellow, shading to orange-brown or reddish-brown towards the base.

Flesh Hard; pale yellow or pale orange; above the tubes and under the surface of the stipe it is lemon-yellow, brownish at the base of the stipe, turns bluish when cut. It has a mild flavour and smells like fresh pine needles.

Spore deposit Olive-brown.

Spores $8\text{-}11 \times 3\text{-}4\mu m$, elliptical and spindle-shaped; smooth; pale yellow.

Habitat It grows from June to November in woodland, alone or in small groups. It is most common on sandy soils.

Use Though it is edible, it does not have much flavour. The young mushrooms are good for pickling.

Remarks It cannot be confused with any other mushroom.

DOWNY BOLETUS, THE GOAT'S LIP

Edible

Xerocomus subtomentosus (L. ex Fr.) Quél.
Syn.: *Boletus subtomentosus* L. ex Fr.

Cap It varies in colour from yellow, yellowish-brown, olive-yellow, olive-brown to greyish-brown; soft and velvety or felt-like, dry and matt. In dry weather the cuticle may crack in a random pattern, revealing the flesh. When young it is hemispherical, later convex; when mature it is cushion-like; 40-110mm (1½-4¼in) in diameter.

The tubes when young are chrome-yellow, when old they are greenish-yellow to olive-brown, 5-15mm (3/16-⅝in) long; when young attached to or running a little way down the stipe, later detached from the stipe. Pores when young are bright yellow, later greenish-yellow; when mature they are yellowish-brown or olive-brown; large and angular; older mushrooms sometimes turn blue when bruised.

Stipe Cylindrical, usually thinner at the base, solid; 60-100mm (2¼-4in) long, 15-20mm (⅝-¾in) thick with a ribbed yellow to reddish-brown surface.

Flesh As soft as butter in the cap and white or creamy; in the stipe it is hard, fibrous and yellow; in damp conditions it turns slightly blue. It has a mild flavour and a fruity smell.

Spore deposit Olive-brown.

Spores 12-14 × 5-6μm, elliptical and spindle-shaped; smooth; honey-yellow.

Habitat It is one of the earliest species of Boletus, growing from the end of May to October in deciduous and leafy woods, most common in oak woods.

Use This excellent, edible mushroom can be prepared in many ways, and is also suitable for drying. It is advisable to peel off the cuticle before cooking because on mature specimens it is slightly bitter.

Remarks A very similar mushroom is the rarer *Xerocomus spadiceus* (Fr.) Quél., which has a brownish-red cap and a rough reticulated stipe. It grows throughout the summer in woodland and is edible.

BAY BOLETUS **Edible**

Xerocomus badius (Fr.) Kühn. ex Gilb.
Syn.: *Boletus badius* Fr.

Cap When young it is chestnut-brown, or dark brown, when mature it is often reddish-brown; smooth, at first soft and velvety, later bare, in damp weather slimy, otherwise dry. The cuticle can be peeled in small pieces off the older mushrooms. Young mushrooms are hemispherical, becoming convex and finally cushion-like; 50-150mm (2-6in) in diameter.

The tubes are at first pale cream, then pale yellow; when mature they are olive-yellow or greenish, turning blue when bruised; 10-20mm (3/8-3/4in) long, attached to or detached from the stipe. Pores when young are minute, off-white or pale yellow; when mature they are bigger and greenish-yellow to olive-green, turning blue when bruised.

Stipe Pot-bellied when young, later cylindrical and solid; 40-120mm (1½-4¾in) long, 8-40mm (5/16-1½in) thick, olive-yellow; streaked with brown fibres or brown all over; it sometimes turns blue when bruised.

Flesh Compact and hard, creamy to pale yellow, turning blue in places when fresh. It has a mild flavour and a typical mushroom smell.

Spore deposit Olive to olive-brown.

Spores 12-16 × 5-6μm, elliptical and spindle-shaped; smooth; honey-yellow.

Habitat This is perhaps the most common tasty *Boletus*. It grows from June to the end of November in mixed woodland especially in spruce woods. The main season is October and November.

Use This excellent, edible mushroom is as good as *Boletus edulis* (The Penny Bun or Cep). It can be cooked in all ways, dried and pickled.

Remarks *Xerocomus badius* cannot be confused with any other *Boletus*.

THE PENNY BUN or CEP **Edible**

Boletus edulis Bull. ex Fr.

Cap Almost white when young, it turns brown to dark brown as it matures; smooth or wrinkled, at first minutely hairy then bare. When young it is hemispherical, later becoming convex then flat or cushion-like; 60-250mm (2¼-10in) in diameter.

The tubes are white, later pale yellow and then greenish-yellow to olive-green; 8-30mm (5/16-1⅛in) long, separated from the stipe. The pores are the same colour; they are usually minute and rounded, and retain their colour when bruised.

Stipe At first pot-bellied or barrel-shaped, later cylindrical or club-shaped; solid; 50-200mm (2-8in) long, 15-60mm (⅝-2¼in) thick. Its surface is usually white or pale cream with a network of fine white fibres covering the upper section of the stipe. Its base is white and felt-like.

Flesh Compact, hard and white; it retains its colour when cut. It has a pleasant flavour and a typical mushroom smell.

Spore deposit Olive-brown.

Spores 14-20×4-6μm, spindle-shaped; smooth; honey-yellow.

Habitat It grows from August to November; it is most common in mountainous spruce woods, but it can also be found in low-lying deciduous woods.

Use It is one of the best mushrooms, and can be cooked in many ways. It is often pickled when young. When dried properly it remains cream-coloured.

Remarks This mushroom is not likely to be confused with any *Boletus* other than the very bitter, inedible *Tylopilus felleus* (Bull. ex Fr.) P. Karst.

SUMMER BOLETUS **Edible**

Boletus aestivalis (Paul.) ex Fr.
Syn.: *Boletus reticulatus* Schaeff. ex Boud.

Cap When young it is grey-yellowish, later leather-brown to light brown; it may be smooth or wrinkled; at first covered as if with fine felt, then bare, dry and matt; in dry weather the cuticle often cracks and shrinks. Hemispherical, then convex, becoming cushion-like at maturity; 60-300mm (2¼-12in) in diameter.

The tubes are white, then yellow-green, 10-35mm (⅜-1⅜in) long, at first notched where they join the stipe then almost detached from it. The pores are rounded and minute; when young they are white, soon they go greenish-yellow; when mature they are olive-green; they do not change colour when bruised.

Stipe When young it is barrel-shaped or pot-bellied; when mature it is club-shaped or cylindrical; solid; 100-200mm (4-8in) long, 20-70mm (¾-2¾in) thick, pale cinnamon or light brown; the surface is covered with a network of fibres, white at the top grading downward to brown. The base is white and felt-like.

Flesh Thick, fleshy and compact; firm when young, soft when mature; white; retains its colour when cut; tastes delicious, and has the typical mushroom smell.

Spore deposit Olive-brown.

Spores 13-20×4-6μm, spindle-shaped; smooth; honey-yellow.

Habitat It grows from the second half of May to the end of June, and again from the end of August to October in oak woods, rarely in coniferous woods; it grows singly or in small groups.

Use This delicious edible mushroom tastes like the Penny Bun, and its smell is even more pleasant. It can be cooked in any way. It needs careful cleaning because it is often attacked by insect larvae.

Remarks A very similar mushroom is *Boletus betulicolus* (Vasil'k.) Pil. et Dermek, which has a paler cap and stipe, with the reticulated surface covering only about one third of the stipe. It grows in summer under birches. It is edible and tasty.

BRONZE BOLETUS **Edible**

Boletus aereus Bull. ex Fr.

Cap Chocolate-brown to very dark brown; smooth, rarely wrinkled; when young it is velvety, dry and matt; when mature it is bare. At first it is hemispherical, later convex and at maturity cushion-like; 50-200mm (2-8in) in diameter.

The young tubes are white, becoming greenish-yellow with a golden brown tinge; 5-25mm ($^{3}/_{16}$-1in) long; notched where they join the stipe, or detached from it. The pores are rounded, and minute; when young they are almost shining white with an uneven surface; later they are olive-yellow tinged with brown.

Stipe When young it is barrel-shaped or pot-bellied; when mature it is usually cylindrical but may be club-shaped; solid, 50-120mm (2-4¾in) long, 20-40mm (¾-1½in) thick; usually a little lighter than the cap, but it may be dark brown. The surface is a fine fibrous network, white at the top, grading downward to brown.

Flesh Compact, firm and white; it retains its colour when cut. It has a pleasant flavour and the usual mushroom smell.

Spore deposit Olive-brown.

Spores 11-17 × 4-6μm, spindle-shaped; smooth; honey-coloured.

Habitat It grows from the end of May to October in deciduous woods, mainly under oaks, singly or in small groups. This rather rare mushroom is more common in warm lowland regions, where it grows under chestnut-trees.

Use This excellent edible mushroom is suitable for any kind of cooking; it tastes as good as the previous three mushrooms.

Remarks It may be confused with *Xerocomus badius* which has no fibrous network on its stipe. Its flesh sometimes turns blue. *X. badius* is also a good, edible mushroom.

BOLETUS CALOPUS **Inedible**

Boletus calopus Fr.

Cap Light brown, olive to greyish-brown; smooth, rarely wrinkled; when young velvety, matt and dry; when mature it is bare. At first hemispherical, it becomes convex with a down-curved irregularly wavy margin; 45-150mm (1¾-6in) in diameter.

The tubes are at first lemon-yellow, later olive-yellow, turning blue when cut; 3-16mm (⅛-⅔in) long; notched where they join the stipe or detached from it. The pores are rounded and minute; greyish-yellow when young, later lemon-yellow, developing a greenish shade; they turn blue when bruised.

Stipe When young it is pot-bellied, later club-shaped or cylindrical, sometimes with a pointed base; solid; 30-150mm (1⅛-6in) long, 10-45mm (⅜-1¾in) thick. Its top is lemon-yellow and finely reticulated; its middle part is carmine with a network of coarser red fibres; its lower part is usually brownish-red to dark brown; the base is white and felt-like.

Flesh Compact, firm, white to pale cream; when cut it turns blue in places, mainly in the cap and at the top of the stipe. It tastes sweet when young but it becomes very bitter; it has little odour.

Spore deposit Brownish-olive.

Spores 12-16×4-6μm, elliptical and spindle-shaped; smooth; orange-yellow.

Habitat It grows from July to October, singly or in small clumps; it is common in coniferous woods, rare in deciduous woods.

Use This beautiful mushroom is inedible because of its unpleasant flavour.

Remarks It may be confused with the poisonous Devil's Boletus, *Boletus satanas* Lenz, but the Devil's Boletus has a white or greyish cap and carmine pores. Similar to *Boletus calopus* is *Boletus radicans* Pers. ex Fr., which grows in deciduous woods and has a stipe which is not red. It is also bitter and inedible.

×

BOLETUS APPENDICULATUS **Edible**

Boletus appendiculatus (Schaeff. ex Fr.) Secr.

Cap Yellow-brown, reddish-brown to dark brown and smooth; at first velvety and matt; later bare and streaked with slender fibrils. It changes from hemispherical through convex to a cushion-like shape; 70-200mm (2¾-8in) in diameter.

The tubes are at first yellow, later olive-green, turning blue when cut; 15-30mm (⅝-1⅛in) long and not attached to the stipe. The pores are round and minute, changing from golden-yellow, to golden-brown and showing greenish-blue spots when bruised.

Stipe At first it is pot-bellied, later cylindrical with a conical base; 50-150mm (2-6in) long, 15-50mm (⅝-2in) thick; solid; deep yellow with a reddish-brown base; the upper half is usually covered by a network of fine white fibres, turning brown when mature.

Flesh Thick, fleshy and firm; yellow except at the base where it is brown or pinkish-brown; it turns light blue in the cap, especially above the tubes. It tastes pleasant and has little odour.

Spore deposit Olive-brown.

Spores 11-15 × 4-6μm, elliptical and spindle-shaped; smooth; honey-coloured.

Habitat It grows from June to September in warm deciduous woods, and is most common under oaks and beeches on chalk; solitary and scattered.

Use This is a very good mushroom, suitable for many ways of cooking. Dried slices are an attractive yellow.

Caution A very similar mushroom is *Boletus subappendiculatus* which is very rare, growing in mountain regions under spruces and firs. It has white flesh with a pleasant flavour and is edible.

ROYAL BOLETUS

Edible

Boletus regius Krombh.

Cap Pink, streaked with slender blood-red fibrils so that the resulting colour is pink to deep carmine red; when ripe it is usually paler. When young it is hemispherical, later convex; when mature it is cushion-like; smooth or very rarely finely wrinkled and dry; at first velvety and matt, finally bare; 60-150mm (2¼-6in) in diameter.

The tubes are lemon-yellow, becoming deep yellow; 10-25mm (⅜-1in) long, then yellowish-green with an olive tinge, not turning blue when cut; notched where they join the stipe, or detached from it. The pores are rounded and minute; at first deep yellow with an olive tinge; when mature their surface is uneven; they do not turn blue when bruised.

Stipe Pot-bellied at first, becoming club-shaped or cylindrical; 50-150mm (2-6in) long, 15-60mm (⅜-2¼in) thick and solid. Its surface is chrome-yellow, the top is covered with a network of fine white fibres. Around its base are blotches which are usually wine-red. When bruised it may turn blue, especially in damp weather.

Flesh Thick, fleshy, pale or deep yellow except at the base of the stipe where it is reddish-brown; it retains its colour when cut. It has little smell but tastes good.

Spore deposit Olive-brown.

Spores 15-17 × 4-5μm, cylindrical to spindle-shaped; smooth; yellowish-brown.

Habitat It grows from the end of May to the second half of September in deciduous woods, mainly under oaks. It is most common on chalky soils in warm regions.

Use A very good edible mushroom which tastes rather like *Boletus appendiculatus*.

Remarks A very similar mushroom is *Boletus speciosus* Frost, that differs from it by having a red stipe and flesh which turns blue. *B. speciosus* grows in summer in warm deciduous woods, mainly under beeches. It is edible and tasty but very rare.

BOLETUS ERYTHROPUS **Edible**

Boletus erythropus (Fr. ex Fr.) Krombh.

Cap Usually dark brown to almost black, rarely paler. At first it is velvety, matt and dry; when mature it is bare; it turns black when bruised. When young it is hemispherical, then convex, becoming cushion-like when mature. 50-200mm (2-8in) in diameter.

The tubes are at first yellow, later greenish-yellow; 10-30mm (⅜-1⅛in) long, notched where they join the stipe, or detached from it. The cap turns blue when cut. The pores are rounded and minute; at first yellow, then orange, soon becoming blood-red; they turn blue when bruised.

Stipe Pot-bellied at first, then club-shaped or cylindrical; solid; 50-150mm (2-6) long, 20-40mm (¾-1½in) thick; the surface is yellow, speckled with light or dark red grainy scales; it turns blue when bruised.

Flesh Thick, firm and deep yellow; when cut it immediately turns a deep greenish-blue; it has little smell or flavour.

Spore deposit Olive-brown.

Spores 13-18×5-5.5μm, elliptical and spindle-shaped; smooth; olive-yellow.

Habitat It grows from June to November in mountainous spruce woods. It is rare under deciduous trees in lower regions. People hesitate to pick it for cooking because it turns blue.

Use This edible mushroom has quite a good flavour but it must be well cooked, otherwise it may upset delicate digestions. It is used for sauces and in meat dishes.

Caution A very similar mushroom is *Boletus queletii* Schulz. This differs in the reddish-brown or wine-red colour of its cap, the orange pores and the almost bare stipe which is usually wine-red in its lower half. The flesh in the cap and the upper half of the stipe is carmine. When cut the yellow parts of the flesh turn blue. It grows from May to October in deciduous woods. It is inedible and quite rare.

LURID BOLETUS **Edible**

Boletus luridus Schaeff. ex Fr.

Cap The colour ranges from olive-yellow to golden orange, sometimes entirely red; at first it is velvety, matt and dry, gradually going bare until mature; it turns blue when bruised but the blue spots soon change to dark brown; when young it is hemispherical, later convex, when mature it is cushion-like; 55-200mm (2⅛-8in) in diameter.

The tubes change from yellow through greenish-yellow to olive, turning blue when cut; 15-35mm (⅝-1⅜in) long, notched where they join the stipe, or detached from it. The pores are rounded and minute; when young they are yellow, then orange to brick-red, turning blue when bruised.

Stipe At first it is pot-bellied, later club-shaped or cylindrical; solid; 40-150mm (1½-6in) long, 20-50mm (¾-2in) thick; its deep yellow surface is covered with a striking network of red fibres which appear to be stretched vertically. Near the base the stipe is wine-red or very dark-brown.

Flesh Thick, fleshy and compact; yellow in the cap, becoming light red in the upper part of the stipe to wine-red at the base. When cut the yellow areas immediately turn greenish-blue.

Spore deposit Olive-brown.

Spores $11\text{-}16 \times 5.5\text{-}6.5\mu m$, elliptical and spindle-shaped; smooth; honey-coloured.

Habitat It grows from the end of May to the beginning of November in deciduous lowland woods and in hilly regions, mainly under oak trees.

Use This edible mushroom can be used in sauces and is suitable for drying. It needs to be thoroughly cooked to avoid stomach trouble.

Remarks It may be confused with some poisonous species such as *Boletus rhodoxanthus* (Krombh.) Kallenb. In contrast with these poisonous fungi, *Boletus luridus* has wine-red or brownish-purple flesh in the lower half of the stipe, not yellow.

DEVIL'S BOLETUS, SATAN'S MUSHROOM

Poisonous

Boletus satanas Lenz

Cap Off-white or very pale grey when young; turning brownish-yellow when older or bruised; sometimes it has a slight green tinge but never any trace of red or pink. When young it is hemispherical, later convex with an inward-curved margin, becoming cushion-like when mature; 60-250mm (2¼-10in) in diameter. The cuticle is dry; when young it is matt and velvety; when mature it is usually bare.

The tubes are at first yellow and very short, later greenish-yellow, turning blue when cut; 10-25mm (⅜-1in) long, notched where they join the stipe, or detached from it. The pores are rounded and minute; when young they are yellow or orange, then carmine to purple at maturity, often with a yellow or orange stripe near the margin of the cap. The pores turn blue when bruised.

Stipe When young it is egg-shaped or spherical, later barrel-shaped, narrowing toward the top; rarely club-shaped or cylindrical; solid; 40-120mm (1½-4¾in) long, 40-100mm (1½-4in) thick. The upper section is usually yellow, the middle is red and the lower section is a dull yellow or brown. The top half is covered with a very fine network of fibres.

Flesh Thick, fleshy and compact; off-white, turning very slightly blue when cut, mainly when young. When young it has little smell, but when mature it stinks like rotting onions.

Spore deposit Olive-brown.

Spores 11-15 × 5-7μm, elliptical and spindle-shaped; smooth; honey-coloured.

Habitat It grows from July to the end of September in open deciduous woods on chalky soil, usually under oak and beech. It is very rare but more often found in warm regions.

Use Poisonous, but only when raw. If well cooked it can be eaten without danger, but it does not taste at all pleasant. Its poisonous quality is often overestimated; there are many other gill-bearing mushrooms that are much more dangerous.

Remarks It is often confused with various species of fungi similar to *Boletus rhodoxanthus*; all are poisonous and quite rare.

†

PURPLE BOLETUS **Poisonous**

Boletus rhodoxanthus (Krombh.) Kallenb.

Cap Orange-yellow to pale yellowish-brown, often with light red or pink tinges. At first it is velvety dry and matt; when mature it is smooth and a slightly darker yellow, turning brown when bruised. When young it is hemispherical, later convex and cushion-like at maturity; 60-200mm (2¼-8in) in diameter. The young tubes are yellow, later greenish-yellow, turning blue when cut; 15-25mm (⅝-1in) long and not attached to the stipe. The pores are rounded and minute, yellow when young, carmine or purple when mature, turning blue when bruised.

Stipe At first it is egg-shaped, later club-shaped or cylindrical; solid; 50-120mm (2-4¾in) long, 30-50mm (1⅛-2in) thick; the deep yellow surface is covered with a network of very fine fibres. When bruised it turns blue.

Flesh Thick, fleshy and compact, deep yellow, turning only slightly blue when cut, mainly when young.

Spore deposit Olive-brown.

Spores $10\text{-}15 \times 4\text{-}5\mu m$, elliptical and spindle-shaped; smooth; olive-yellow.

Habitat It grows from the end of July to mid-September, usually on chalky soils in warm deciduous woods under oak trees. It is quite rare but can be found more often than the Devil's Boletus, *Boletus satanas.*

Use When raw it is poisonous. Some people eat it cooked with no ill-effects, but in spite of that it cannot be recommended for consumption.

Remarks Several other fungi are related and similar to *Boletus rhodoxanthus.* They differ from it in the colour of the cap, stipe and flesh; for example the stipe of *Boletus splendidus* Martin is not completely covered by net and its pale yellow flesh turns a brighter blue; *Boletus rhodopurpureus* (Smotl.) is all carmine and its deep yellow flesh turns dark blue. Both are poisonous.

†

HORNBEAM BOLETUS **Edible**

Leccinum griseum (Quél) Sing.
Syn.: *Leccinum carpini* (R. Schulz) Moser

Cap Yellowish-brown through grey-brown to almost black; bare, with a wrinkled, uneven surface; in dry weather the cuticle usually cracks and shrinks, revealing the flesh. When young it is hemispherical, later convex; when mature it is cushion-like; 80-140mm (3⅛-5½in) diameter.

The tubes are at first off-white or cream, later yellowish-grey, becoming a drab olive-yellow at maturity. They turn greyish-purple to very dark grey when cut. The pores are rounded and minute with an uneven surface; when young they are off-white or cream, darkening to yellowish-grey.

Stipe Cylindrical, often thickened at the base, narrowing under the cap; solid; 50-130mm (2-5⅛in) long, 10-35mm (⅜-1⅜in) thick; longitudinally wrinkled at the top. On its off-white or cream-coloured surface it may have minute dark brown scales.

Flesh Quite soft in the cap but fibrous in the stipe, off-white to pale yellow; when cut it first turns greyish-purple, then goes completely black. It has a mild flavour and little smell.

Spore deposit Tobacco-brown.

Spores 10-20 × 5-7μm, spindle-shaped; smooth; brownish-yellow.

Habitat It grows from June to the end of October in deciduous woods, mainly under hazels and hornbeams.

Use It is edible and tastes much like other species of *Leccinum*. Unfortunately, it is often attacked by insect larvae. It is suitable for sauces and stews.

Remarks A very similar mushroom is *Leccinum duriusculum* (Schulzer ap. Fr.) Sing: its cap is greyish- to reddish-brown, its tubes and pores are at first off-white; when mature they are a pale greyish cream; the stipe has a pale background covered with very tiny brown scales. The flesh is usually thick, fleshy, very firm and white; when cut the cap turns pink and the stipe turns blue but never black. *L. duriusculum* grows in summer and autumn, only under white poplar trees. It is edible and very tasty.

WHITE POPLAR BOLETUS **Edible**

Leccinum duriusculum (Schulzer ap. Fr.) Sing.

Cap Reddish-brown, yellowish-brown or greyish-brown; when young it is streaked with tiny scales or fine hairs; when mature it is bare, dry and matt; at first hemispherical, becoming convex then cushion-like; 60-150mm (2¼-6in) in diameter.

The tubes are 12-25mm (½-1in) long, white or off-white, then greyish-cream or pale grey, taking on a brown tinge when bruised.

Stipe 50-150mm (2-6in) long, 10-30mm (⅜-1⅛in) thick, cylindrical or spindle-shaped; the upper section is white or creamy, the lower section is brownish; the base has a blue tinge and is covered with very fine brown scales.

Flesh White, shading to salmon pink in the upper section of the stipe and blue in the lower section. The flesh of the whole mushroom is firm; it has a mild flavour and a typically pleasant mushroom smell.

Spore deposit Olive-brown.

Spores $13\text{-}17 \times 7\text{-}5\mu m$, elliptical and spindle-shaped; smooth; honey-coloured.

Habitat It grows in and near woods from the end of July to mid-November, singly or in small groups under various species of poplar, usually near the white poplar.

Use It is perhaps the best *Leccinum* to eat because its firm, fine-flavoured flesh is seldom spoilt by insects. It can be served in various ways but is particularly good in meat dishes and is suitable for drying.

COW FUNGUS
BROWN BIRCH BOLETUS

Edible

Leccinum scabrum (Bull. ex Fr.) S. F. Gray
Syn.: *Boletus leucophareus* Pers.

Cap Brown, in a range of shades from yellowish to very dark; smooth and bare; in damp weather slightly sticky, in dry weather slightly shiny. When young it is hemispherical or bell-shaped and convex; when mature it is cushion-like; 50-150mm (2-6in) in diameter. The tubes are at first off-white, later greyish-cream to pale grey; 10-25mm (⅜-1in) long; detached or partly joined to the stipe. They retain their colour when cut. The pores are rounded and minute; when young they are pale cream; when mature they are yellowish-grey.

Stipe Cylindrical, tapering slightly towards the top and thickening; a little thicker at the base; solid; 80-170mm (3⅛-6⅝in) long, 10-35mm (⅜-1⅜in) thick, with coarse brown to dark-brown scales arranged longitudinally on a pale grey background.

Flesh White or pale cream, retaining its colour when cut. The whole young fruiting body is pliable and firm; when mature it is soft and watery in the cap but fibrous and tough in the stipe; it has a mild flavour and little smell.

Spore deposit Brown.

Spores 14-20×4-6μm, spindle-shaped; smooth; yellowish-brown.

Habitat It grows from July to the end of November in deciduous woods under birches, singly or in small groups.

Remarks There are several similar mushrooms, including the White Birch Boletus, *Leccinum holopus* (Rostk.) Watling, and Black Birch Boletus, *Leccinum melaneum* (Smotl.) Pil et Dermek. *L. holopus* grows in peat-bogs under birches. It has a white cap and pale cream tubes and pores; its young stipe is entirely white, turning brown and scaly. Its flesh is white, keeping its colour when cut. It is edible but very rare. *L. melaneum* grows in damp soil under birches. Its cap is almost black; the stipe has a creamy surface with black scales; its white flesh retains its colour when cut. It is edible and locally common.

ORANGE BIRCH BOLETUS **Edible**

Leccinum testaceoscabrum (Secr.) ex Sing.

Cap Orange or golden-yellow when young, turning pale and less orange when mature; at first velvety, matt and dry, later bare. The young cap is hemispherical, with the margin touching the stipe, then convex and finally cushion-like; 60-160mm (2¼-6¼in) in diameter.

The tubes are pale grey to olive-grey, 10-30mm (⅜-1⅛in) long, detached or partly joined to the stipe. The pores are rounded and minute, at first grey tinged with olive, ripening to a dusky brown.

Stipe Club-shaped or cylindrical; solid; 80-220mm (3⅛-8⅝in) long, 20-70mm (¾-2¾in) thick, with black scales on a white or greyish surface. When mature the scales turn brown.

Flesh Thick, fleshy and compact; white, turning pinkish- or greyish-purple when cut; it shades to bluish-green in the base of the stipe. The whole young mushroom is pliable and firm; when mature it is soft in the cap but fibrous in the stipe. It tastes mild and pleasant, with little smell.

Spore deposit Brown.

Spores 13-17 × 4-6μm, spindle-shaped; smooth; honey-coloured.

Habitat It grows from June to the end of October under birches.

Use This good edible mushroom is suitable for stewing and for flavouring meat dishes. It is popular because it is very fleshy and rarely damaged by insects.

Remarks A very similar fungus to *Leccinum decipiens* (Sing.) Pil. et Dermek. It grows in autumn under white poplars. The cap is yellowish-brown or dark brown; the stipe is brown and scaly; the flesh is firm and white, turning greyish-purple when cut; at the base of the stipe it turns blue. It is edible and tasty.

ORANGE-CAP BOLETUS

Edible

Leccinum aurantiacum (Bull. ex St-Am.) S. F. Gray

Cap Reddish-orange, dark red or brick red, when young it is somewhat velvety, matt and dry; bare when mature. At first it is hemispherical with the margin touching the stipe, becoming convex and eventually cushion-like; 40-150mm (1½-6in) in diameter. The margin of the cap is thin; the cuticle at the edge usually covers the tubes slightly like a thin veil.

The tubes are at first off-white, later pale grey with an olive tinge; 7-30mm (3/10-1⅛in) long; notched where they join the stipe or detached from it. The pores are rounded and minute; cream when young, olive-grey when mature.

Stipe Cylindrical or club-shaped; solid, 50-180mm (2-7in) long, 15-50mm (⅝-2in) thick; at first covered with white scales which turn rusty-brown or chestnut.

Flesh Thick fleshy, white and firm; when cut it turns first greyish-purple then brownish-red and finally dark grey. It is good to eat but has little smell.

Spore deposit Brown to olive-brown; when dried it is tobacco-brown.

Spores 13-17 × 4-5μm, spindle-like; smooth; yellowish-brown.

Habitat It grows from July to November under aspens. It is easy to spot because of its red cap.

Use This tasty mushroom is best stewed in meat dishes and risotto. It dries well, turning dark grey.

Remarks *Leccinum quercinum* Pil. is very similar. It differs from *L. auranticum* in its reddish-brown cap and stipe; the stipe has brown scales which are most tightly packed towards the middle. The mushrooms are almost always big and firm. They grow from June to November in oak woods and are very good to eat.

BITTER BOLETUS **Inedible**

Tylopilus felleus (Bull. ex Fr.) P. Karst.
Syn.: *Boletus felleus* Bull.

Cap Pale leather-brown, yellowish-brown or dark brown and smooth; when young it is velvety, matt and dry; when mature it is bare. It starts hemispherical, becoming convex and finally cushion-like; usually centred on the stipe but sometimes lop-sided; 50-120mm (2-4¾in) in diameter. The tubes are at first white, then greyish-pink; when mature they are flesh-pink, 15-20mm (⅝-¾in) long; detached or partly joined to the stipe. The pores are slightly angular and minute; when young they are white, later pinkish; when mature they are flesh-pink, turning brown when bruised.

Stipe At first it is pot-bellied, later club-shaped or cylindrical, symmetrical or lop-sided; solid; 50-125mm (2-5in) long, 10-25mm (⅜-1in) thick; usually white or cream at the top; elsewhere it is off-white or orange-yellow; covered with a network of brown fibres which stand out from the surface of the stipe.

Flesh Pliable and white; it retains its colour when cut. The taste is sweet initially, becoming bitter and sharp; it has little smell.

Spore deposit Rose to flesh-pink.

Spores 10-15×4-6μm, elliptical and spindle-shaped; smooth; colourless or a very pale greyish-pink.

Habitat It grows from June to the end of October, usually in coniferous woods, sometimes in deciduous woods. It may be found singly or in groups. It is most common in pine woods on sandy soil.

Use Inedible because it is extremely bitter. If you add only a few slices of this mushroom to a dish, they spoil the food completely.

Remarks *Tylopilus felleus* may be confused with *Boletus aestivalis* (Paul.) ex Fr. At first glance the two are very similar. But *T. felleus* when mature has flesh-pink tubes and pores and coarse brown network on its stipe. *B. aestivalis* has greenish-yellow tubes and pores and its stipe has delicate white network. The best way to distinguish the two species when they are young and their pores are still white is to taste them.

SAFFRON MILK CAP **Edible**

Lactarius deliciosus (L. ex Fr.) S. F. Gray

Cap Yellowish-orange or brick-orange, with rusty-brown concentric rings; it often has greenish blotches. When young it is convex with the margin turned inward; later it is flat, depressed in the centre or funnel-shaped; 50-150mm (2-6in) in diameter. The cuticle is smooth, at first velvety, later bare.

The rather closely-set gills extend down the stipe; 4-6mm (about ¼ in) wide, golden orange when young, deep orange when mature, turning green when bruised.

Stipe Cylindrical, 30-70mm (1⅛-2¾in) long, 10-25mm (⅜-1in) thick; spongy at first and hollow when mature. Its surface is smooth or dimpled and orange or reddish-orange with greenish areas.

Flesh Very brittle, white or creamy-yellow; when cut it immediately turns orange to carrot-red, later it goes paler and greenish. It smells pleasant and tastes spicy and piquant. Its 'milk' is carrot-coloured.

Spore deposit Creamy-yellow.

Spores $7\text{-}9 \times 6\text{-}7\mu m$, spherical or egg-shaped with short, close-set spines linked in a pale cream or almost colourless network.

Habitat It grows from August to November in mountainous spruce woods, profusely if the weather is good.

Use Saffron Milk Cap is a delicious, edible mushroom, the most tasty *Lactarius*. Recommended ways of serving it include stewing it with onions and paprika and adding it to an omelette. It is good pickled.

Remarks *Lactarius deliciosus* var. pini Vasilk grows in lowland pine woods. It is more fleshy and firmer; the rings on its cap are clearer and it is usually bigger than *L. deliciosus*. It is as tasty but rarer.

SHAGGY MILK CAP
WOOLLY MILK CAP

Inedible

Lactarius torminosus (Schaeff. ex Fr.) S. F. Gray

Cap Flesh-pink to rose-red; at first covered in a thick velvety or shaggy coat; later almost bare except at the margin. When young it is convex; when mature it has a navel-like depression in the centre and a margin curving inward considerably; 50-120mm (2-4¾in) in diameter.

The gills run a little way down the stipe; they are very close together, 3-4mm (about ⅛in) thick; at first pale cream, later pinkish-yellow.

Stipe Cylindrical and central; 30-70mm (1⅛-2¾in) long, 10-25mm (⅜-1in) thick; spongy at first, becoming hollow when mature; smooth or covered with dimples; yellowish-red or rose.

Flesh Firm, brittle and white; it retains its colour when cut. It has a faint smell, sometimes a little like turpentine. Its 'milk' is white and acrid.

Spore deposit Pale cream.

Spores 7.5-10×6-7.5μm, spherical or egg-shaped with blunt projections on its surface joined in a fine network; colourless.

Habitat It grows from September to the end of October under birches, singly or in quite large clumps.

Use It is inedible because of its strongly acrid taste. However, some people stew it in water until it loses its acrid taste then throw away the liquid. After this treatment it can be pickled or preserved in salt.

Remarks A very similar mushroom is *Lactarius pubescens* Fr. It has fainter rings on its cap and is paler. It grows in summer and autumn in deciduous woods. It also is inedible.

LACTARIUS VOLEMUS **Edible**

Lactarius volemus (Fr.) Fr.

Cap Yellowish-brown to orange-brown with no concentric rings; dry and matt; at first it is velvety; when mature it is smooth and often cracked. It is convex when young, with a margin which curves slightly inward; later it becomes almost flat but depressed in the centre; at maturity it is funnel-shaped; 80-150mm (3⅛-6in) in diameter.

The closely-set gills extend down the stem; 4-7mm ($^{1}/_{6}$-$^{3}/_{10}$in) wide; at first creamy, then yellowish, turning brown when bruised.

Stipe Cylindrical and solid; 50-120mm (2-4¾in) long, 15-25mm (⅝-1in) thick, with a smooth bare surface; yellowish- or reddish-brown but always paler than the cap.

Flesh Firm and brittle, creamy to pale yellow, smelling of herring. Its 'milk' is abundant; white, turning grey in fresh air. At first it is quite sweet, after a while it becomes unpleasantly acrid.

Spore deposit Creamy to pale orange-yellow.

Spores 8-12 × 7-11 μm, almost spherical, with fine spines linked in a network on the surface; colourless.

Habitat It grows from June to October in mixed woodland, singly or in small groups but never in large colonies. It is most common in hilly country.

Use This mushroom can be eaten raw with a little salt, or fried in butter with caraway seeds — it is an acquired taste.

Remarks *Lactarius volemus* cannot readily be confused with any inedible or slightly poisonous *Lactarius*. One slightly poisonous species is *L. helvus* (Fr.). Its cap is yellowish-brown with a reddish tinge and no rings. Its gills are at first off-white, then orange-yellow becoming orange when mature. The stipe is yellowish-brown and when mature it is hollow. Its flesh is pale orange-yellow, smelling like chicory. Its 'milk' is colourless and quite bitter. It grows in mixed woodland. Large amounts may cause slight poisoning.

BARED-TOOTHED RUSSULA **Edible**

Russula vesca Fr.

Cap Pale orange-yellow, flesh pink or brownish-red, often with pink or purple spots on a pale yellow surface; the margin is smooth, later finely grooved. When young it is hemispherical, then convex and at maturity it is flat but depressed in the centre; 40-120mm (1½-4¾in) in diameter. The cuticle is bare and quite easily peeled.

The gills are fragile, white when young, developing rusty spots at the edge when mature; closely set; 7-10mm (3/10-3/8in) wide; not attached to the stipe; rounded near the margin of the cap.

Stipe Cylindrical, narrowing at the base; solid; 30-60mm (1⅛-2¼in) long, 15-25mm (⅝-1in) thick; firm, with a white surface but yellowish near the base.

Flesh Fragile, white, turning rusty-brown when cut. It has a mild flavour like hazelnuts. When fresh it has little odour; when old it smells like herring.

Spore deposit Chalk-white.

Spores 6-8×5-6μm, almost spherical or egg-shaped, with scattered dots on the surface; colourless.

Habitat It grows from June to October in mixed woodland. It is plentiful in good weather but unfortunately it is often attacked by insects.

Use This edible mushroom is among the most tasty species of *Russula*. It is suitable for soups and stews.

Remarks The slightly poisonous Sickener, *Russula emetica* (Schaeff. ex Fr.) S. F. Gray also has a pink or red cap, but differs from *R. vesca* in having smaller fruiting bodies with creamy-yellow gills and a sharp, hot taste. Some people may be made slightly unwell if they eat it.

GREEN AGARIC, BLUE AND YELLOW RUSSULA

Edible

Russula cyanoxantha (Schaeff. ex Schw.) Fr.

Cap Grey, in a range of shades; usually grey-green, slate-grey or purple, with a yellow to orange-yellow centre and sometimes a pink margin. In damp weather the cuticle is sticky and slimy, streaked with slender fibrils. When young the cap is hemispherical, later convex; when mature it is flat with a depressed centre; 50-160mm (2-6¼in) in diameter.

The gills are soft, pliable, firm and closely packed; 5-10mm (3/16-⅜in) wide; not attached to the stipe and rounded near the margin; at first they are white, turning creamy-yellow.

Stipe Cylindrical, at first solid and firm, later spongy and brittle; 50-120mm (2-4¾in) long, 15-30mm (⅝-1⅛in) thick; the surface is often wrinkled, bare and white, sometimes with pale purple shading.

Flesh Firm, flexible and juicy; white, retaining its colour when cut; it has a mild flavour, rather like nuts, but little smell.

Spore deposit Chalk-white.

Spores 8-9 × 7-8μm, almost spherical with scattered dots on its surface; colourless.

Habitat It grows from June to November in mixed woodland at low and high altitudes.

Use This delicious mushroom is perhaps the most tasty *Russula*. It is used in soups and meat dishes and is pickled with other kinds of mushrooms. It goes well with *R. virescens* and *R. vesca* as an accompaniment to stews.

Remarks A similar mushroom is *Russula palumbina* Quél. Its cap is slate-grey, olive-grey or purplish-grey; its gills are at first white, then creamy and easily broken; the stipe is white tinged with pink or purple and may be wrinkled. The flesh is white with a mild flavour and little smell. It grows in summer and autumn in deciduous woods, mainly under beeches. It is edible.

GRASS-GREEN RUSSULA **Edible**

Russula aeruginea Lindbl. in Fr.
Syn.: *Russula graminicolor* (Secr.) Quél. s. auct.

Cap At first it is hemispherical, when mature it is quite flat, often with a depression in the centre; 30-100mm ($1\frac{1}{8}$-4in) wide, with a grooved margin; in shades of grass-green but usually darker in the centre. The cuticle when young is slimy or sticky and can be fairly easily peeled.

The gills are at first white, later straw-yellow; 8-10mm ($\frac{5}{16}$-$\frac{3}{8}$in) wide; closely-set, thin, fragile, fully attached to the stipe and rounded near the margin of the cap.

Stipe Cylindrical, 40-80mm ($1\frac{1}{2}$-$3\frac{1}{8}$in) long; 10-20mm ($\frac{3}{8}$-$\frac{3}{4}$in) thick; when young it is solid; when mature it becomes white and spongy; sometimes it has rusty spots at the base; smooth, bare and silky.

Flesh Easily broken, white, retaining its colour when cut; it tastes hot but has little smell.

Spore deposit Rich cream.

Spores 8×6-$7\mu m$, almost spherical, with fine spines on the surface forming an incomplete network; colourless.

Habitat It grows from June to October in mixed woodland, mainly under spruce and birch.

Use Though its flesh and gills taste sharp when raw, it is edible. It loses this tart flavour when cooked.

Remarks *Russula heterophylla* (Fr.) Fr. is sometimes similar to *R. aeruginea*. Its cap is usually yellowish- or olive-green with a brown centre. The gills are white at first, becoming pale yellow, often with a fork-like split near the cap margin. It grows from July to October in deciduous woods, mainly under beeches. It is edible.

YELLOW SWAMP RUSSULA **Edible**

Russula claroflava Grove s. Melz. et Zv.
Syn.: *Russula flava* (Rom.) Rom. ap. Lindbl.

Cap 50-80mm (2-3⅛in) wide; when young it is hemispherical or convex, later flat; bare and matt; lemon-yellow or chrome-yellow; the cuticle can be peeled fairly easily.

The gills are 6-8mm (¼-5/16in) wide; close-set and not attached to the stipe; white when young, later butter-yellow; when old or scratched they turn dark grey.

Stipe 40-60mm (1½-2¼in) long, 8-12mm (5/16-½in) thick; cylindrical; at first smooth, then finely wrinkled; when young it is white, when mature it becomes grey; at first solid then spongy.

Flesh Very fragile; white, turning grey when cut; it tastes mild or slightly bitter, with little smell.

Spore deposit Cream to pale orange-yellow.

Spores 8-10×7-8μm, elliptical with spines on the surface forming an incomplete network; colourless.

Habitat It grows in damp, grassy deciduous woods, especially under birch, poplar and alder, from July to October in small groups.

Use It is edible but not very tasty.

Remarks A similar mushroom is *Russula decolorans* (Fr.) Fr. It is bigger and has a brick-red or orange cap. It grows from July to the end of September in damp, hilly pine woods. It is edible and quite tasty.

RUSSULA OLIVACEA **Edible**

Russula olivacea (Schaeff. ex Secr.) Fr.

Cap Purple, wine-red, violet, olive-green or orange-yellow, either a single colour or mixed; it bears concentric rings of more intense shade. At first it is bare, later velvety. When young it is almost spherical, becoming convex; flat; when mature it is often depressed in the centre, with a smooth margin; 100-200mm (4-8in) in diameter.

The gills are 6-10mm (¼-⅜in) wide, close-together and rather thick; at first they are partly joined to the stipe, then detached; rounded near the margin of the cap; when young they are yellow, turning more orange as they mature.

Stipe Cylindrical or spindle-shaped, 60-150mm (2¼-6in) long, 25-40mm (1-1½in) thick; smooth and velvety; creamy with a purple tinge, often all carmine; when mature it is yellow to rusty-brown at the base. When young it is solid, later becoming spongy.

Flesh When young it is juicy and white; retaining its colour when cut; it has a mild flavour and little smell.

Spore deposit Deep yellow.

Spores 8-13 × 8-12μm, almost spherical, with isolated spines on the surface; pale yellow.

Habitat It grows from June to October in mixed woodland, often thriving in dry weather. It is more frequent in hill country then in lowlands.

Use It is a useful mushroom, with solid flesh, appearing in early summer before other species. It can be cooked like *Russula cyanoxantha*.

Remarks *R. alutacea* (Pers. ex Fr.) Fr. is similar in colour and size. Its cap may be wine-red, brownish-red, purple, olive, yellowish-green or orange-yellow; the gills turn from creamy-yellow to a deep yellow; the stipe is solid, cylindrical, and bright white. Its flesh is white and it smells pleasant. It grows from June to September in deciduous woods. It is edible and tasty.

GOLDEN RUSSULA **Edible**

Russula aurata (With.) ex Fr.

Cap In young specimens it is vermilion to orange-red, developing red and orange blotches on a yellow background or turning chrome-yellow all over. When young it is almost spherical, later convex; when mature it is flat and wide, often depressed in the centre and with a grooved margin; 60-120mm (2¼-4¾in) in diameter. The cuticle is at first bare, slightly slimy and shiny, later velvety and matt.

The gills are 6-10mm (¼-⅜in) wide, moderately packed together and not attached to the stipe; rounded near the margin; at first they are creamy, later butter-yellow with chrome-yellow edges; when mature they are very brittle.

Stipe Cylindrical or club-shaped, 35-80mm (1⅜-3⅛in) long, 15-25mm (⅝-1in) thick; white tinged with yellow, smooth or slightly wrinkled, usually bare. When young it is hollow, when mature the interior becomes spongy.

Flesh Very brittle; off-white but golden yellow beneath the cuticle of the cap, retaining its colour when cut. It has a mild flavour and little smell.

Spore deposit Yellow.

Spores 9-12 × 9-11 μm, almost spherical; the surface bears warts joined in a honeycomb network; yellowish.

Habitat It grows from June to the end of September in mixed woodland. Singly, never in groups.

Use This tasty mushroom has one disadvantage — its flesh is very fragile.

Remarks At first sight *R. lepida* (Fr.) looks similar, but its flesh is firm and its cap is always vermilion. When raw its flesh tastes mild and smells like fruit, but when cooked it smells like turpentine and is therefore inedible. It grows from July to October in mixed woodland.

RUSSULA XERAMPELINA **Edible**

Russula xerampelina (Schaeff. ex Secr.) Fr.

Cap The colour varies greatly according to the kind of trees under which it grows. In coniferous woods it has a shining carmine, wine-red or purple cap; in beech woods it is paler, usually pink, olive or brownish-red; under birches it is yellow or yellowish-green, sometimes with a purple tinge. When young the cap is hemispherical, later convex and at maturity it is flat with a depression in its centre; 60-120mm (2¼-4¾in) in diameter.

The gills are at first creamy, later yellowish-cream, turning brown when bruised; 7-12mm (3/10-½in) wide; closely packed, brittle and not attached to the stipe; rounded near the margin of the cap.

Stipe Cylindrical, 40-80mm (1½-3⅛in) long, 15-30mm (⅜-1⅛in) thick; bare, smooth or finely wrinkled; white, with a red tinge or often all red. Inside it is solid, becoming spongy and brittle.

Flesh At first it is white or creamy; when cut it turns brown in places; when young it tastes slightly sharp with little smell; later the flavour is milder but it smells like herring.

Spore deposit Pale orange-yellow.

Spores 8-13 × 8-11 μm, broadly elliptical with spines on the surface; creamy-yellow.

Habitat It grows from August to the end of November in mixed woodland scattered in small groups.

Use It is edible and of average quality. It is generally eaten with more tasty mushrooms.

Remarks *Russula melliolens* (Quél.) looks like the red version of *R. xerampelina*. Its cap is red or reddish-brown; the gills are white at first, turning creamy-yellow; the stipe is white, sometimes slightly reddish. The flesh when young is white, later showing orange-yellow spots. It grows in summer in deciduous woods, mainly under oak trees. It smells pleasantly like honey and is edible.

COMMON PUFF-BALL, POOR MAN'S SWEETBREAD

Edible

Lycoperdon perlatum Pers. ex Pers.
Syn.: *Lycoperdon gemmatum*

Fruiting body Usually club-shaped or pear-shaped. Its spherical spore-producing part is 20-50mm (¾-2in) in diameter; the cylindrical lower section is sterile and usually 20-60mm (¾-2¼in) long, 12-22mm (½-⅞in) thick. The young mushroom is all white, bearing spiny warts which are mainly concentrated over the spherical area; when mature it is orange-yellow to brown and bare. Inside it is white and pliable when young. When it ripens it dries up and the white interior changes into olive-brown spore powder which escapes through an opening in the top.

Spores 3.5-4.5μm, spherical, with fine warts on the surface; pale orange-yellow.

Habitat It grows from June to November in mixed woodland, almost always in large groups. It is very common.

Use It is good to eat while the interior is still white and pliable, but the rubbery cuticle must be removed before cooking.

Remarks *Calvatia excipuliformis* (Scop. ex Pers.) Pers. is very similar. Its fruiting body is 50-150mm (2-6in) tall, club-shaped or pear-shaped; the spherical upper part is 30-80mm (1⅛-3⅛in) wide. When young it is white, the upper part covered with pointed warts; later it is bare and yellow. When ripe it is brown. There is no hole in the top but the upper part simply falls to pieces leaving just the stipe. The inside is at first white and pliable; when mature it changes into dark brown spore powder. The spores are 5-6μm across, spherical; dark brown with spines on their surface. This puff-ball grows from July to November in mixed woodland and grassy woodland margins. It is edible and can be cooked the same way as *Lycoperdon perlatum*.

COMMON EARTHBALL **Poisonous**

Scleroderma citrinum Pers. (*S. aurantium* or *S. vulgare*)

Fruiting body Bulbous or nearly so, 30-100mm (1⅛-4in) across with a yellow to orange-yellow surface. The upper area is cracked and covered with thick warts; the lower section is usually wrinkled, bare and slightly narrowed, with a root-like bunch of mycelium fibres. The outer wall is very thick (2-4mm, about ⅛in). The spore-producing interior is white at first; when mature it is black and interwoven with white sterile fibrils. When ripe the inside changes into olive-brown spore powder and the top splits open in several places.

Spores 7-15μm, spherical with a network of spines on the surface; dark brown.

Habitat It grows from July to November in mixed woodland, usually on sandy soil.

Use Poisonous, but only in large amounts. It is harmless if only 2-3 slices are added to a dish. It is eaten because its taste and smell are a bit like some species of edible truffles.

Remarks Small mushrooms with brown shiny caps grow as parasites on the Common Earthball. This is a clump of *Xerocomus parasiticus* (Bull. ex Fr.) Quél, which is edible but extremely rare.

†

Calendar of Edible Mushrooms in season

Name	Page	Months											
		1	2	3	4	5	6	7	8	9	10	11	12
Bared-toothed Russula	196						●	●	●	●	●		
Bay Boletus	156						●	●	●	●	●	●	
Blewits	74										●	●	●
Blusher	90							●	●	●	●	●	
Boletus appendiculatus	166						●	●	●	●			
Boletus erythropus	170						●	●	●	●	●	●	
Bronze Boletus	162					●	●	●	●	●	●		
Brown Birch Boletus	182							●	●	●	●	●	
Brown Wood Mushroom	110						●	●	●	●	●	●	
Cauliflower Fungus	46								●	●	●		
Changeable Agaric	126				●	●	●	●	●	●	●	●	●
Chantarelle	48					●	●	●	●	●	●	●	
Chestnut Boletus	140							●	●	●	●	●	
Clouded Agaric	68									●	●	●	●
Common Morel	40			●	●	●							
Conical Morel	42			●	●	●							
Cow Boletus	150						●	●	●	●	●	●	
Deer Mushroom	100				●	●	●	●	●	●	●	●	●
Dingy Agaric	78										●	●	●
Downy Boletus	154					●	●	●	●	●	●		
Fairy Ring Mushroom	84					●	●	●	●	●	●	●	
Field Mushroom	112					●	●	●	●	●	●	●	
Golden Russula	208						●	●	●	●			

Name	Page	Months											
		1	2	3	4	5	6	7	8	9	10	11	12
Gomphidius glutinosus	134							•	•	•	•	•	
Gomphidius rutilus	136							•	•	•	•	•	
Grass-green Russula	202						•	•	•	•	•		
Green Agaric	198						•	•	•	•	•	•	
Green Cracking Russula	200						•	•	•	•	•		
Gypsy Mushroom	124							•	•	•	•		
Honey Fungus	64									•	•		
Horn of Plenty	50								•	•			
Hornbeam Boletus	178						•	•	•	•	•		
Horse Mushroom	108						•	•	•	•	•		
Indigo Boletus	138							•	•	•	•		
Lactarius volemus	194						•	•	•	•	•		
Larch Boletus	142						•	•	•	•	•	•	
Lurid Boletus	172					•	•	•	•	•	•	•	
Lyophyllum decastes	60									•	•	•	•
Man on Horseback	76										•	•	•
March Mushroom	58			•	•	•							
Orange Birch Boletus	184						•	•	•	•	•		
Orange-cap Boletus	186							•	•	•	•	•	
Oyster Mushroom	56	•	•	•							•	•	•
Parasol Mushroom	104							•	•	•	•		
Penny Bun	158								•	•	•	•	
Puff-Ball	212						•	•	•	•	•	•	
Rose-gilled Grisette	102					•	•						
Royal Boletus	168					•	•	•	•	•			
Russula olivacea	206						•	•	•	•	•		
Russula xerampelina	210								•	•	•	•	
Saffron Milk Cap	190								•	•	•	•	
St. George's Mushroom	62				•	•	•						
Scaly Hydnum	52								•	•	•	•	
Shaggy Cap	114				•	•	•	•	•	•	•	•	
Shaggy Parasol	106							•	•	•	•		

Name	Page	Months											
		1	2	3	4	5	6	7	8	9	10	11	12
Sheathed Agaric	86						●	●	●	●	●		
Slippery Jack	144					●	●	●	●	●	●	●	●
Suillus granulatus	148						●	●	●	●	●	●	
Suillus placidus	146						●	●	●	●	●	●	
Summer Boletus	160								●	●	●		
Sweetbread Mushroom	128							●	●	●	●	●	
Trumpet Agaric	66								●	●	●	●	
Variegated Boletus	152						●	●	●	●	●	●	
White Poplar Boletus	180							●	●	●	●	●	
Winter Mushroom	82										●	●	●
Wood Blewits	72									●	●	●	●
Wood Hedgehog	54							●	●	●	●	●	
Yellow Swamp Russula	204							●	●	●	●		

Index of English Names

Index of Latin Names